Atomic Absorption Spectrometry: An Introduction

Atomic Absorption Spectrometry: An Introduction, 2nd Edition

ALFREDO SANZ-MEDEL AND ROSARIO PEREIRO

MOMENTUM PRESS

MOMENTUM PRESS, LLC, NEW YORK

Atomic Absorption Spectrometry: An Introduction, 2nd Edition
Copyright © Momentum Press®, LLC, 2014.

First published by Momentum Press®, LLC
222 East 46th Street, New York, NY 10017
www.momentumpress.net

ISBN-13: 978-1-60650-435-2 (hardback, case bound)
ISBN-13: 978-1-60650-437-6 (e-book)

DOI: 10.5643/9781606504376

Cover design by Jonathan Pennell
Interior design by Exeter Premedia Services Private Ltd., Chennai, India

10 9 8 7 6 5 4 3 2 1

Printed in the United States of America

Contents

Preface

More than 50 years have elapsed since the successful introduction of atomic absorption spectrometry (AAS) by Sir Alan Walsh for elemental analysis. At this moment in time it is comparatively rare to find research publications on new AAS fundamentals or even new AAS applications. Conversely, analytical AAS has become an established elemental analysis routine tool applied worldwide. Furthermore, it can be safely said that in today's routine laboratories AAS is the most popular, inexpensive, and easy-to-use technique among all the "workhorses" available for elemental analysis.

It is not surprising then that we have restricted the scope of this tutorial book to the study of fundamentals and practical use of such popular and efficient atomic absorption techniques. An up-to-date account of AAS fundamentals, instrumentation, special techniques, and elemental analysis applications is provided here. To do so, the atomic absorption experiment and the photophysical law governing such photon absorption processes are revised first. Then, the main components or units, that, when adequately assembled, constitute an AAS instrument, are described in detail to set the foundations of modern spectrometers for AAS measurements.

After describing the basic instrumentation for AAS measurements, we have selected those AAS-based techniques that are not only well established now, but of recognized practical value to provide robust and fit-to-purpose analytical results. In order to provide direct, concise material, we utilized the approach of studying progressively complex AAS techniques that have been introduced to solve problems of lack of sensitivity, selectivity, or easy sample handling in elemental determinations, as carried out worldwide today in routine laboratories. A section of carefully selected applications (illustrating the relative merits and strengths of each particular AAS technique) is included at the end of each chapter.

The approach of using a classical flame as atomizer is considered first, while the following chapter deals with flame sensitivity problems and how to overcome them in a general way by resorting to electrothermal (ET) atomization of the sample to be analyzed. The greater sensitivity of ET, coming from a confined and better-controlled atomization of the whole sample (not just the analyte), explains the superior performance of ETAAS in ultratrace determinations (μg/L concentrations of the analyte in the dissolved sample) versus mg/L in conventional flame AAS (FAAS).

The next chapter deals with a rather special chemical, a more restricted alternative to improve the sensitivity of those analytes able to form volatile species at room temperature when the dissolved sample is treated with appropriate reagents; for instance, techniques based on hydride or cold-vapor generation to increase analyte transport. Of course, such higher sensitivity approaches so popular today in AAS are not restricted to atomic absorption (even if they were born and developed mainly using absorption measurements).

Also discussed is flow injection analysis (FIA), a practical technique intended for the automation of serial assays, which has evolved well beyond that original point. At present, FIA offers the most convenient and straightforward approach to not only automate, but also to enhance, all sample pretreatment steps for atomic spectrometric detectors in general and for atomic absorption techniques in particular. Also, the online coupling of chromatographic separations to AAS for elemental detection has demonstrated a great analytical potential, in particular for metal speciation purposes. Therefore, a complete section is dedicated to such couplings and problem solving based on chemical speciation.

The final chapter is devoted to emerging fields of applications (such as the characterization of nanoparticles) and to chemometrics. A section focused on troubleshooting and quality-control guidelines has also been included.

The book ends with:

- Appendix A—Buyer's Guide: a listing of manufacturers worldwide of atomic absorption spectrometers and information about companies marketing AAS-based instruments;
- Appendix B—Glossary: definitions of specialized terms;
- Appendix C—Standards: selected examples of standards for chemical analysis of different samples by AAS (from the British Standards Institution and the International Standards Organization) are provided; and
- References: A full list of the research articles, review articles, and books that were cited in each chapter.

KEYWORDS

atomic absorption spectrometry (AAS); characterization of nanoparticles; chemometrics; chromatographic separations; cold vapor generation; electro-thermal (ET) atomization; elemental analysis; flow injection analysis (FIA); hydride vapor generation

1

An Introduction to Analytical Atomic Spectrometry

In this introductory chapter, the basic interactions of ultraviolet–visible photons with atoms (spontaneous emission, stimulated absorption, and stimulated emission) are reviewed using self-explanatory diagrams. Furthermore, the important characteristics of the atomic lines, constituting the atomic spectra, are considered in detail paying particular attention to main effects and phenomena in an atomizer that gives rise to the broadening of atomic lines and brings about the final spectral width of an atomic line.

It ends with a critical comparison of the main analytical techniques based on atomic spectrometry (absorption, emission, fluorescence, and mass spectrometry) for the analysis of dissolved samples as well as for the direct analysis of solids.

1.1. BASIC INTERACTIONS OF ELECTROMAGNETIC RADIATION WITH ATOMS FOR CHEMICAL ANALYSIS

Most of the molecular spectroscopy–based analytical methods currently used take advantage of the differential features of interacting photons of radiation with the matter to be analyzed. In particular, the interaction of photons with electrons of the molecules of matter is observed and measured. The photons most frequently used for such analytical purposes extend from the ultraviolet (UV: 190–390 nm) to the visible (Vis: 390–750 nm) regions of the electromagnetic spectrum; these interact easily with valence electrons. Of course, the infrared (IR: 750–2,500 nm) region is also very useful in molecular analysis.

Three main basic photon–electron interactions of the molecules are analyzed: absorption, emission, and fluorescence. If the energy of photons, $E = h \cdot v$, is equal to the energy gap between electronic energy states 1 and 2 ($\Delta E = E_2 - E_1 = h \cdot v$), the so-called resonance condition is fulfilled and the photon disappears as its energy is absorbed to promote the electron to the higher-energy state 2 (excited state of the electron). Furthermore, the electron may be promoted to an excited state by other types of energy (thermal, electric, etc.). Once excited, the electron is not stable at the higher energy level and tends to fall back to the lower energy (nonexcited) level. In doing so, the energy difference $\Delta E = E_2 - E_1$ can be emitted as electromagnetic radiation of frequency $v = \Delta E/h$ (h is the Planck's constant: 6.63×10^{-34} J s). When the energy used to promote the electron to the excited state is in the form of photons (photoexcitation), the spontaneous emission occurring is called fluorescence.

This tutorial book deals with analytical atomic spectrometry. In this case, the analyzed matter (sample) must be in the form of atoms in the gas phase, but the main interactions we make use of for the analysis are the same: that is, absorption, emission, and fluorescence. From an analytical point of view, the fact of having the sought elements of the sample in the form of gaseous atoms is most convenient because we obtain atomic spectra. Such spectra (graphs of the measured absorption, emission, or fluorescence versus the energy of the photons used in the experiment) are consequently much simpler to interpret than molecular spectra.

The interaction processes between UV–Vis photons and outer electrons of the atoms of the sought elements can be studied and understood through the quantum mechanics theory. In the thermodynamic equilibrium between matter and interacting electromagnetic radiation, according to the radiation laws postulated by Einstein, three basic processes (quantized transitions) between the two stable energy levels 1 and 2 are possible. These three processes, which can be defined by their corresponding transition probabilities, are summarized in Figure 1.1:

(a) *Spontaneous emission of photons.* This process refers to a spontaneous transition of the electron from the excited state 2 to the lower energy state 1 with emission of a photon of frequency $v_0 = (E_2 - E_1)/h$. The probability of such a transition is usually represented by A_{21} (and corresponds to the inverse value of the lifetime of the excited state 2).

(b) *Stimulated absorption of photons.* In this case, the electronic transition takes place from state 1 to state 2 in response to the action of an external radiation of the appropriate frequency. The probability of such transitions can be represented by B_{12}.

(c) *Stimulated emission of photons.* This process consists of electronic transitions from the excited energy level to the lower one stimulated or in response to an external radiation of the adequate frequency $(E_2 - E_1)/h$. The probability of these transitions is termed B_{21} (of course, a given transition only has a given probability and so $B_{21} = B_{12}$).

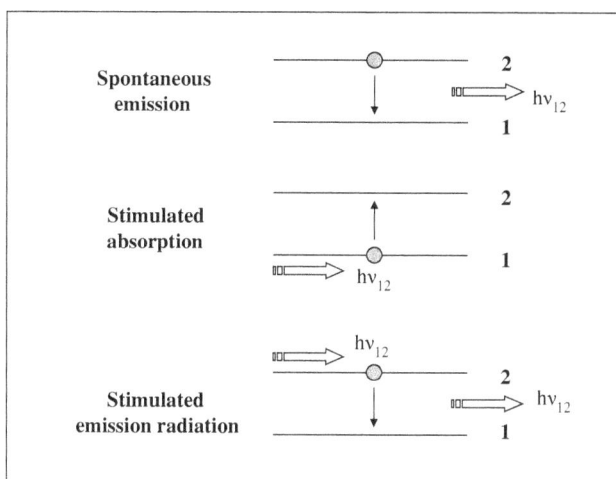

Figure 1.1. Basic interaction processes between matter
and interacting electromagnetic radiation.

The first process (a) constitutes the photophysical basis of atomic emission spectrometry, while atomic absorption spectrometry (AAS) is based on the second one (b). Fluorescence is the sequential combination of a stimulated absorption (b) followed by spontaneous emission (a).

The third process (c) constitutes the basis of the light amplification by stimulated emission of radiation (LASER) phenomenon, which is opposite to the absorption process (contradicting the common misconception that emission and absorption are really opposite phenomena). In atomic spectrometry, the LASER process has been and is being used for atomization, ionization, or to build spectrochemical sources or lamps of very specific and advantageous features. In fact, the application of laser sources in this field is considerable, but the basic interaction taking place in the atomizer itself, where the sample is located for the corresponding analysis, is not a laser process (as stated before, the analytically useful basic interactions are absorption, emission, and fluorescence).

Atomic spectra are usually a series of very narrow peaks (e.g., a few picometers bandwidth) whose analytical beauty consists of: first, the observed peak frequency, $v = \Delta E/h$, will tell us the element present (because the electronic transition energy is characteristic of the element considered); second, at a given frequency (peak), we are looking at a given element, so the measured peak height or area will inform us about the concentration of that element in the sample. The relative simplicity of such atomic spectra and their straightforward qualitative and quantitative information have led to the widespread practical importance and present predominance of atomic spectrometry for inorganic elemental analysis.

However, it should be stressed that to gain such simplicity of spectra we have to transform the elements constitutive of the sample into active species, that is, into atoms that will interact with the photons of the UV–Vis radiation. This requirement brings about at least one practical and one fundamental consequence:

- From a practical point of view we need an atomizer, usually a dynamic medium of high temperature where molecules of the sample are cleaved and broken down into individual gaseous atoms.
- From a fundamental point of view it should always be borne in mind that in the atomizer all molecular information is basically destroyed. The information provided by analytical atomic spectrometry is elemental. If molecular information is required, molecular spectroscopy should be used, or we should resort to a previous separation of the different types of compounds, containing the elements to be measured, present in the sample. Many separation techniques can be directly coupled to the atomic detectors (as carried out in speciation analysis).

1.2. ATOMIC LINE SPECTRA AND THEIR ORIGIN

The transitions of outer electrons of an atom may be represented as vertical lines on an energy-level diagram, as shown in Figure 1.2 where each energy level of the outer electron possesses a given energy and is represented by a horizontal line. The vertical lines represent the energy differences between a given pair of quantized energy levels ($\Delta E = E_n - E_m$) and $E_n - E_m$ is the energy involved in such an electronic transition. Of course, each of those energy levels corresponds to an orbit of the electron around the nucleus of that atom, as illustrated in Figure 1.3.

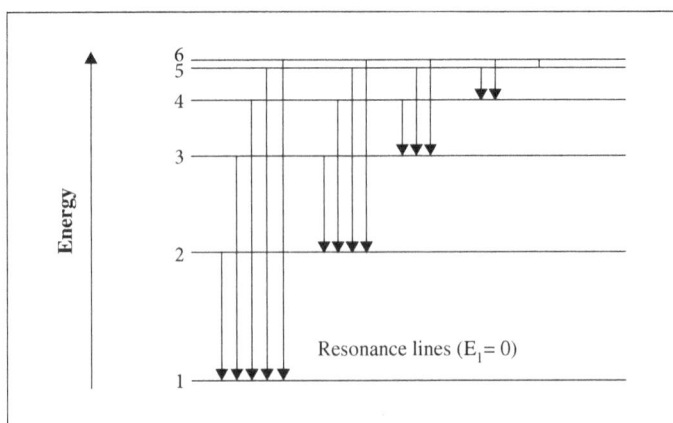

Figure 1.2. Energy levels and electronic transitions.

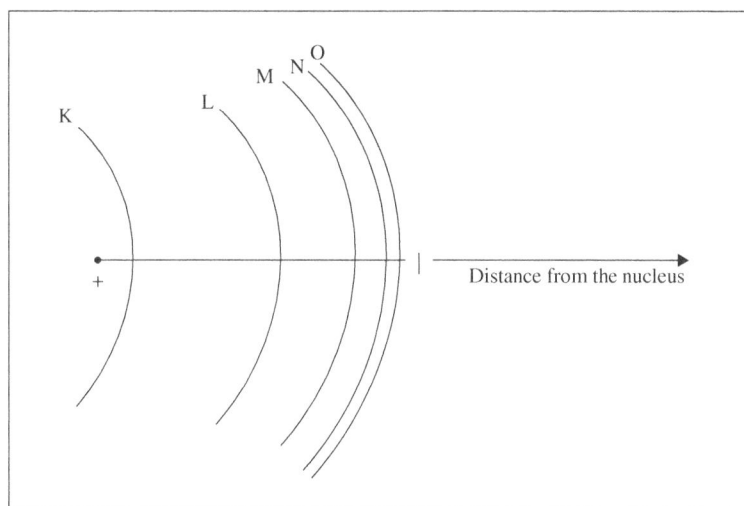

Figure 1.3. Geometric arrangement of electron shells.

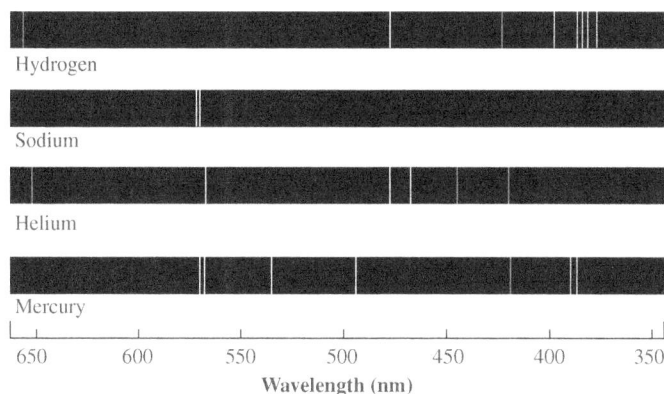

Figure 1.4. Some examples of atomic emission spectra
(photographic plates).

   All electronic transitions ending at the same energy level (i.e., horizontal
line of Figure 1.2) are usually called a series, the most likely ones being those
ending at the lowest possible energy level (the ground state) of the electron
in the atom. Of course, there are transitions observed that have their lower
energy transition level above the ground state.

   Historically, the frequency of the light coming from such transitions (or
its wavelength, $\lambda = c/v$) was separated in a monochromator, and the intensity observed for each frequency was registered on a photographic plate;
this constituted the atomic spectrum. As an example, Figure 1.4 shows such
photographic spectra made of many glowing atomic lines (or transitions) for
hydrogen, sodium, helium, and mercury.

Today the detection of light is electronically based (e.g., with a photomultiplier) rather than photographic. If the observed intensity of the emitted light (radiation) is plotted versus the frequency (or wavelength) of the corresponding line (transition), a typical atomic emission spectrum, made of peaks at the wavelengths of emission, is obtained.

If a stimulated absorption of light in response to an electronic transition between a lower and a higher energy level is measured, a similar plot, percent absorption versus frequency of the light, can be drawn, as illustrated in Figure 1.5a. Such a plot represents a typical atomic absorption spectrum that we will use extensively throughout this book, specifically devoted to this type of measurements in atomic vapors.

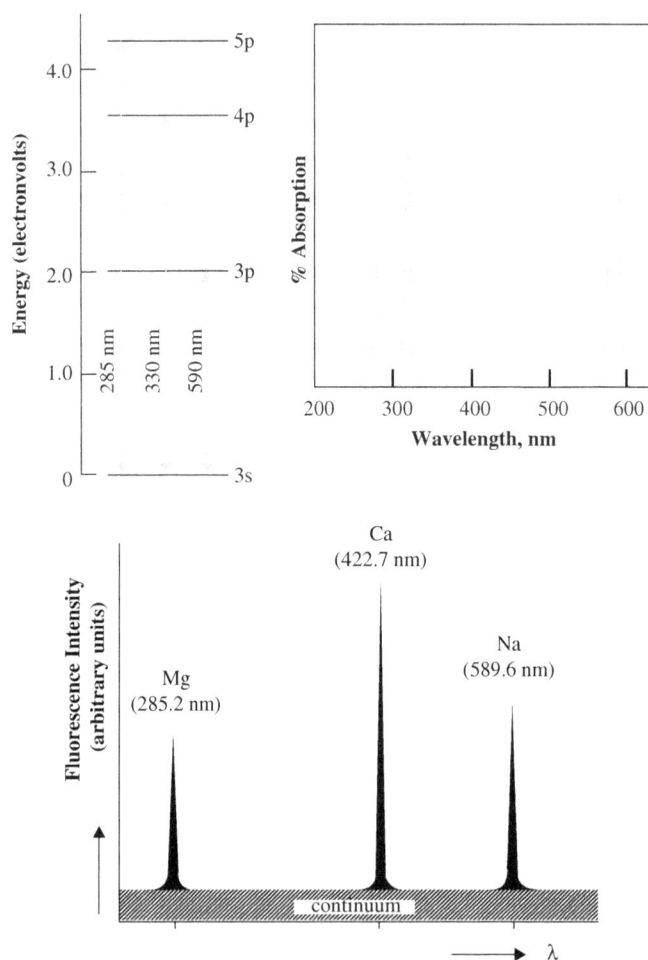

Figure 1.5. Examples of typical atomic absorption and atomic fluorescence spectra. (a) The typical atomic absorption spectrum of sodium vapor and corresponding sodium atom transitions. (b) A typical atomic fluorescence spectrum for main lines of a mixture of Mg, Cd, and Na atoms.

Finally, an atomic fluorescence spectrum would be the plot of the measured fluorescence intensity, coming from a cloud of atoms excited by an appropriate radiation, as a function of the frequency of the emitted radiation (Figure 1.5b).

As shown in Figure 1.5 (atomic spectra registered with electronic detection) the atomic lines are typically represented today as individual peaks at the corresponding wavelength. Such lines carry very precious analytical information: while the frequency v ($v = c/\lambda$) at which the peak occurs corresponds to a given transition (characteristic of a given element), its peak area relates to the concentration of that atom (element) in the sample.

1.3. ATOMIC LINE CHARACTERISTICS

Among all possible transitions (lines of the spectrum), the selection of the optimum atomic line for qualitative and quantitative purposes is critical. The choice is usually the most intense line of the analyte, particularly if it is not interfered with by other lines of concomitant elements in the sample. As a rule, the most intense lines are resonance atomic lines (i.e., when the lowest energy level in the corresponding transition is the fundamental level, $E_1 = 0$). This selection of the atomic (emission, absorption, or fluorescence) line characteristic of the analyte is the first step in quantitative analysis. Measurements of the intensity observed at that line are the basis of the corresponding analyte concentration determination.

As we have seen, the atomic lines in the spectrum appear as vertical lines or peaks (occurring in a very narrow interval of wavelengths) due to the nature of the transition involved. That is, in molecules, an electronic transition (ΔE_{el}) is usually accompanied by simultaneous changes in the molecule vibrational (ΔE_{vib}) and rotational (ΔE_{rot}) energy content; sometimes, all the three energy types may change simultaneously in an electronic transition within a molecule ($\Delta E_{mol} = \Delta E_{el} + \Delta E_{vib} + \Delta E_{rot}$). The many transition possibilities allowed in this way plus the solvent effect derived from the aggregation state of the sample in some techniques (in which the sample is dissolved) determines, for instance, that in UV–Vis molecular absorption (or emission), the corresponding peaks in the spectrum are widely broadened. Typically, the half-bandwidth of an absorption band in molecular UV–Vis spectra is around 40 nm (or 400 angstrom, Å), while in atomic lines, the half-bandwidth observed, resulting from pure electronic transitions, ΔE_{el}, is of a few hundredths of an Å (typically 0.03–0.05 Å).

Spectral interferences in atomic spectroscopy are less likely than in molecular spectroscopy analysis. In other words, one of the strengths of atomic spectroscopy for elemental analysis is its comparatively high freedom from spectral interferences.

However, even the atomic lines are not completely monochromatic (i.e., only one wavelength per transition). There are several phenomena that bring about certain broadening as well. Therefore, any atomic line shows a profile (distribution of intensities measured) as a function of wavelength (or frequency).

The theoretical study of the profile of atomic lines has been a classic in atomic physics; analytical atomic spectrometry has borrowed such theoretical knowledge in order to predict two basic aspects of the calibration function used for quantitative analysis:

(a) The dependence between the intensity measured (emission, absorption, or fluorescence signals) and the analytical concentration sought.
(b) The dependence of the intensity measured with experimental variables and parameters—a most important consideration in order to optimize these parameters and hence achieve good sensitivity for the determinations.

As the analytical selectivity is conditioned by the overall broadening of the lines (particularly the form of the wings of such atomic lines) a basic knowledge of the broadening phenomena is also important.

Therefore, it is important to summarize here the basic concepts and parameters necessary to define in practice the atomic line profile as a function of wavelength (or frequency). From a practical perspective, Figure 1.6 illustrates a typical plot of such a profile observed for an atomic (it could be absorption, emission, or fluorescence) line. As shown in the figure, the main parameters

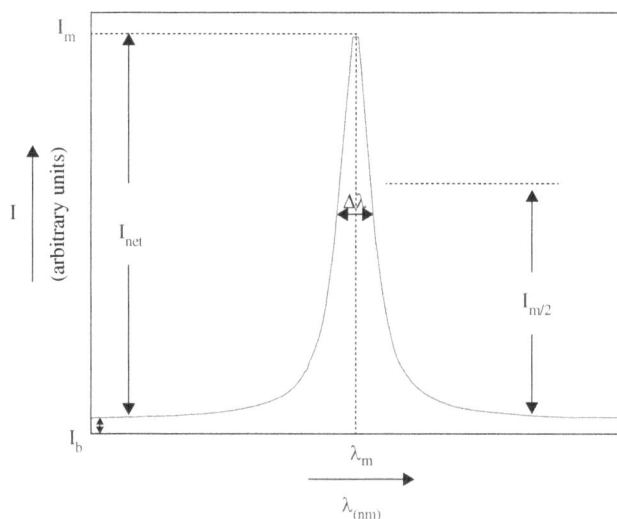

Figure 1.6. A typical atomic line profile.

defining an atomic line are as follows—I_{m}: intensity of the emission, absorption, or fluorescence measured at the line maximum; λ_{m}: wavelength measured at the line maximum; $\Delta\lambda$: half-bandwidth in wavelength units; and I_{b}: intensity measured for the background at λ_{m}.

According to Figure 1.6, the total analytical signal, I_{net} (across the profile), would be given by:

$$I_{\mathrm{net}} = \int_{-\infty}^{+\infty}\left[I(\lambda) - I_{\mathrm{b}}(\lambda)\right]d\lambda$$

(area below the atomic line with corrected background in every point).

At the maximum of the line (peak height): $I_{\mathrm{net}} = I_{\mathrm{m}} - I_{\mathrm{b}}$.

Figure 1.6 shows a typical distribution of intensities around the λ_{m}, showing the broadening of the line around that value. In the following section, the main broadening phenomena bringing about such characteristic profiles of atomic lines will be discussed.

1.4. ATOMIC LINE SPECTRAL WIDTH

Let us illustrate the concept of broadening and the final spectral width of atomic lines, using an absorption line profile for the sake of simplicity. In fact, the concept and theoretical treatment shown below should be similar when describing the profiles and spectral widths of emission (or fluorescence) atomic lines.

As stated before, an atomic absorption line is the result of a pure electronic transition from a lower- to a higher-energy state of the atom stimulated by a photon of adequate frequency. Notwithstanding what was said for Figure 1.2, those absorptions are not infinitely fine vertical lines at frequency $v_0 = \Delta E/\mathrm{h}$.

As measured, the lines exhibit a certain finite width around the maximum absorption wavelength. Atomic lines are also broadened and the extent of such broadening can be measured as the half-width of the line, $\Delta\lambda$. This parameter means the width of the profile, % absorption versus wavelength, at the point at which the maximum absorption is halved.

In order to describe broadening of atomic lines whatever the wavelength considered in the profile, it is preferred in AAS to express the half-width in terms of frequency, Δv (cm^{-1} or s^{-1}).

Thus, as $v(\mathrm{s}^{-1}) = c/\lambda$, after differentiation, $\Delta v(\mathrm{s}^{-1}) = (c/\lambda^2)\,\Delta\lambda$, with $c = 3 \times 10^8$ m/s (and $\Delta v(\mathrm{cm}^{-1}) = \Delta\lambda/\lambda^2$).

Although there are other possible causes for broadening of atomic lines (e.g., interactions of the atoms with electrically charged particles or, at higher concentrations, with one another), the profile of an atomic line is governed mainly by the combined effect of the following processes: natural, Doppler, and Lorentz broadening.

1.4.1 Natural Broadening of Lines

This type of broadening is associated with the degree to which the energy of the levels is defined. By Heisenberg's uncertainty principle, from the point of view of quantum electrodynamics, the energy, E, of a system in a given energy state and the time, t, during which that system remains in such a state cannot be known accurately simultaneously. This is formulated as:

$$\Delta E \cdot \Delta t \geq h/4\pi$$

In our case, this means that if we consider a given atom we must have uncertainty in the time spent by the electron in the excited state before spontaneous deactivation.

In other words, the uncertainty in the finite lifetime, t, of the levels between which a transition takes place determines an actual broadening of the energy levels:

$$\Delta E \geq (h/4\pi)/\Delta t.$$

As the ground level is stable ($\tau = \infty$), it is the width of the upper level which is the only broadening of significance for resonance transitions (see Figure 1.7). Thus, the natural width of a resonance line can be defined as:

$$\Delta v_N = B_{12}/2\pi = 1/2\pi\tau$$

Different spectral lines have different natural widths (e.g., Hg: 253.7 nm, $\Delta v_N = 5.3 \times 10^{-5}$ cm^{-1}; Cd: 228.8, $\Delta v_N = 2.7 \times 10^{-3}$ cm^{-1}), but it is worth

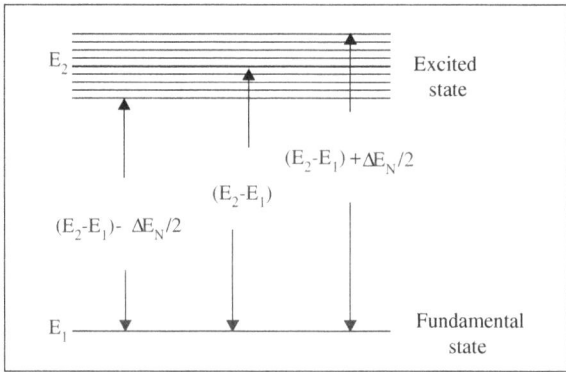

Figure 1.7. Natural line-broadening origins.

noting that in most cases the natural width of atomic lines does not exceed 10^{-3} cm^{-1} and so, as we will see later on, this broadening can then be ignored when considered alongside other broadening phenomena.

1.4.2 Doppler Broadening

The Doppler broadening of atomic lines arises from the random thermal motion of atoms, in the high temperature atomizer, relative to the observer or detector. As for the Doppler effect distorting the frequency of a sound emitted by a moving object (see Figure 1.8), if an electronic transition stimulated by the absorption of a radiation of frequency v_0 takes place in an atom moving at a speed v_x in the line of the observer, the measured frequency of absorption by the atom is displaced by Δv_D, where

$$\Delta v_D = v_0 \cdot (v_x / c)$$

If the atomic vapor in the atomizer (e.g., a flame) is in thermodynamic equilibrium, it can be demonstrated that the motion of atoms there can be described by the Maxwell distribution and then the Doppler half-width of the line is equal to:

$$\Delta v_D = (2v_0 / c)(2 \ln RT / W_{at})^{1/2}$$

(a)

Detector

V$_x$

(b)

Detector

$-V_x$

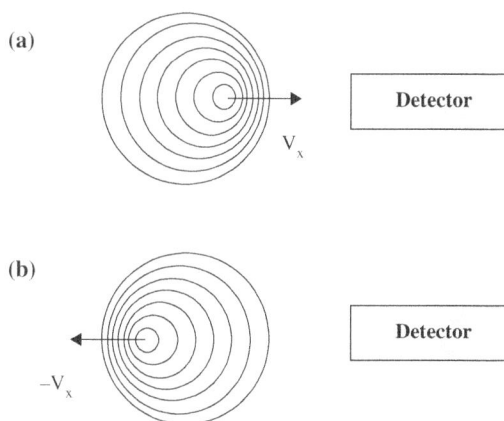

Figure 1.8. Schematic representation of the Doppler effect. As a consequence of the velocity V_x of the atom in the line of sight the observed frequency of absorption (v) by the atom is displaced by Δv. When the atom moves in the opposite direction to the detector, situation (b), v is higher than v_0, while the value of v is less than v_0 for situation (a).

Table 1.1. Calculated values of Doppler broadening (Δv_D) of some atomic lines used for chemical analysis of Cs, Li, and B at two temperatures

Atomic line (Å)	Atomic weight (amu)	Δv_D (cm^{-1}) (500°K)	Δv_D (cm^{-1}) (3000°K)
Cs 8521	132.91	1.6×10^{-2}	4×10^{-2}
Li 6708	6.94	9.1×10^{-2}	2.2×10^{-1}
B 2498	10.81	2×10^{-1}	4.8×10^{-1}

Note: It is important to stress that even at relatively low temperature (500°K) this broadening, Δv_D, is much greater than Δv_N.

and, if numerical values are given to the constants,

$$\Delta v_D = 7.16 \times 10^{-7} \cdot v_0 \, (T/W_{at})^{1/2}$$

(where T is the absolute temperature in K and W_{at} the relative atomic mass of the moving atom).

It is important to note that for a given analyte this broadening is independent of the concentration of the absorbing atoms in the atomizer cell and is proportional to the square root of the absolute temperature.

This Doppler broadening exhibits a Gaussian profile and its value is usually two orders of magnitude higher than natural broadening. Table 1.1 illustrates the calculated values of Δv_D for atomic lines of three chemical elements (atoms) at two different temperatures. Furthermore, this effect is the predominant one defining the upper part of the atomic lines observed in a flame and describes quite accurately the observed profiles above the half intensity of maximum of the lines. Also, this is the main factor determining the experimental width observed in the lines emitted by hollow cathode lamps used for AAS.

1.4.3 Lorentz Broadening

This effect, along with the Doppler phenomenon discussed earlier, makes the predominant contribution to the eventual shape, width, and position of the absorbing lines.

The Lorentz broadening, Δv_L, is observed when the pressure of a foreign gas in the atomizer is increased. Experience shows that collisions with the gas cause broadening, asymmetry of the line profile, and a displacement of the maximum of the line, relative to its initial position. Both, displacement of the original λ_{max} (maximum shift) and line asymmetry, can be easily evaluated in a quantitative manner (see Figure 1.9). It is known that different foreign gases have different effects on the broadening and shifting of the lines.

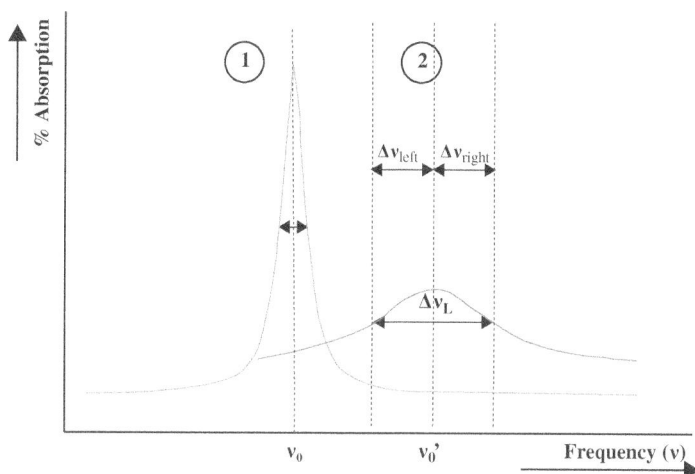

Figure 1.9. Effects of the pressure (Lorentz broadening) on line profiles: 1. Normal profile 2. Lorentz broadening (broadened and shifted profile).

Furthermore, such changes, for a given gas, are proportional to the change in total pressure in the atomizer.

Different theoretical approaches have been proposed to interpret this Lorentzian or pressure broadening. While the collisional theory offers the best fit equations to explain broadening at the center of the line (which is the most important part in atomic absorption, as we will see in the next chapter), the so-called statistical theory describes better these events at the wings of the line.

According to the work of Lorentz, back in 1905, atomic emission of light comes from the harmonic vibration of electrons within the atom. In the presence of a foreign gas, the atoms may collide with the particles of this and so a vibration is interrupted in the collision, while the vibration is renewed at the same frequency immediately afterward.

In the light of this collisional theory, it has been shown that the Lorentz half-width depends on:

$$\Delta v_{L} = 2.6 \times 6.02 \times 10^{23} \cdot \delta^2 \cdot P\,[2/(\pi RT) \cdot (1/W_{at} + 1/W_{mol})]^{1/2}$$

where P is the pressure of the gas, W_{at} the atomic weight of the atom, W_{mol} the molecular weight of the gas, and δ^2 is the effective cross-section of the collision atom–molecule considered.

From a quantum mechanics point of view, the collision would modify the excitation state of the atom, altering in that way the lifetime of such an excited level. As a result, a broadening of the line qualitatively similar to Δv_{N} (but much more intense) would take place.

It should be stressed at this point that the experimental absorption profiles of analyte atoms in the flame (the most common atomic absorption atomizer) are governed by Doppler and Lorentz broadening: the center of the line primarily by the Doppler effect and the wings by the Lorentz effect (see Figure 1.10).

1.4.4 Self-Absorption Effects

In emission measurements, it is well known that at low atom density levels increasing the concentration of the analyte in the atomizer enhances proportionally the height of the observed emission line. However, after a given concentration, the profile of the atomic line broadens rather than increasing its height (see Figure 1.11). This concentration effect is due to self-absorption of emitted radiation by the excess of analyte atoms present in the ground level. As maximum absorption occurs always in the center of the line, this effect is much more pronounced in the center than in the line wings, and the phenomenon will apply particularly to resonant lines (where the lower energy level of the transition is the ground-state level).

Self-absorption effects, although better described for emission measurements, are also present as a cause of broadening in all atomic lines. For instance, in a flame, where all the analyte atoms are roughly at the same temperature, this self-broadening of the lines is mainly responsible for the loss of linearity of calibration curves (absorbance versus concentration) at higher concentrations.

To conclude this section it should be mentioned that when a temperature gradient exists in the atomizer and the absorbing analyte atoms are at lower

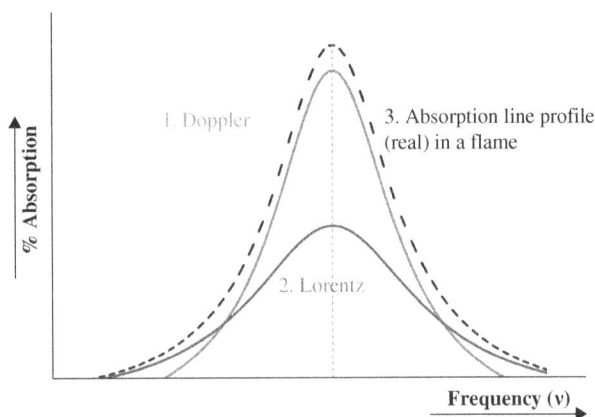

Figure 1.10. A typical symmetrical profile of an absorption line in a flame: The center of the line behaves as Doppler—and the wings are Lorentz—broadened.

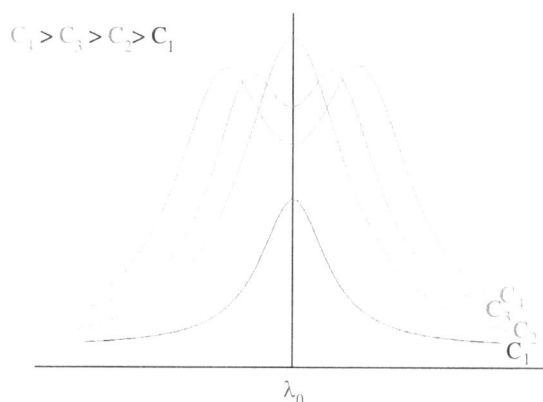

Figure 1.11. Autoabsorption and self-reversal of a spectral line.

temperature than the emitting ones (e.g., in an electric arc discharge), the profile of the absorption is narrower than that of the emission. Thus, strong self-absorption may take place just around λ_{max} and it may be so intense there that the observed atomic line splits into two distinct lines (see Figure 1.11). This is the self-reversal phenomenon, often observed in resonance emission lines in lamps used as spectral light sources at high atomic density emissions, as we will see in Chapter 3 when describing the Smith–Hieftje background corrector.

1.4.5 Other Broadening Processes

Other broadening effects, affecting the profile of atomic lines, have been observed in different atomizers and spectral sources and under different conditions. Although none of those broadening processes is of general practical importance in AAS using a flame, for the sake of completeness, we will mention the following effects:

- Stark broadening that takes place in the presence of (high) electric fields.
- Zeeman broadening occurs when working in a magnetic field. Of course, in flames those processes are of no significance, but Zeeman broadening should be considered also for background correction (see Chapter 3) in electrothermal atomic absorption techniques (as described in Chapter 5).
- Holtsmark broadening is related to Lorentz broadening. Such a process is due to collisions between atoms of the same kind (e.g., the analyte) and thus it is also called resonance broadening in classical books. Of course, as relative concentrations of the analyte are so low, resonance broadening is of negligible importance compared with Lorentz broadening.

Finally, hyperfine structure should be mentioned as it accounts for an apparent broadening of lines, which is in fact due to overlapping of individual lines (because the resolving power of the monochromator used is not good enough). As it is known, interactions between the spin of the nucleus of the atom (different in the presence of several isotopes) and the resultant spin moment of its electrons may result in a splitting of the corresponding energy levels, that is, in a given multiplicity of expected energy levels in that atom (e.g., Hg vapor emission).

An interesting example of isotopic hyperfine structure in atomic spectrometry is the emission of a Hg lamp: the Hg 253.7 nm line is in fact a multiple peak emission resulting from 10 possible excited energy levels (derived from even-numbered Hg isotopes).

1.5. A COMPARATIVE OVERVIEW OF ANALYTICAL ATOMIC SPECTROMETRIC TECHNIQUES

From the preceding discussions it follows that analytical atomic spectrometry is not just one technique. Different techniques based on atomic absorption, atomic emission, or atomic fluorescence are available.

Typically, a dissolved sample is used for analysis to form a liquid spray of the sample that is delivered to the atomizer (e.g., a flame). However, direct solid analysis is also possible by using special atomizer or excitation sources such as arcs, sparks, lasers, or glow discharges. Next, we will see a comparative overview of atomic techniques categorized into two groups, depending on whether the sample is in a liquid or in a solid state.

1.5.1 Dissolved Sample Analysis Techniques

Flame-atomic absorption spectrometry (FAAS), electrothermal atomization (ETAAS), and inductively coupled plasma-optical emission spectrometry (ICP-OES) are probably today's workhorses for dissolved sample routine analysis.

Instrumental development as well as analytical applications (which increased after commercialization of AAS in the 1960s and of ICP-OES in the 1980s) have been profound and extensive during these years. Knowledge about such techniques is now well advanced and so new spectacular break-throughs are not likely. Thus, the capabilities and analytical limitations are well known when comparing their respective analytical performance characteristics. An attempt in that direction is summarized in Table 1.2, which gives a general

Table 1.2. Comparative advantages and limitations of the most common atomic "workhorses" of dissolved sample analysis

	Flame AAS	Electrothermal AAS	ICP-OES
General advantage	Simple and reliable Most widespread Moderate interferences Ideal for unielemental monitoring in small labs High sample throughput	Sub-ppb (μg/L) DLs Microsamples (<1 mL)	Multi-elemental high temperature (7,000°K) Low matrix interferences High dynamic range
Cost	Low cost	Higher cost	High instrument cost
Limitations	Unielemental Sub-ppm (mg/L) DLs* Low temperature (refractory compounds problem) Only metal-metalloids (>60 elements)	Unielemental Not so easy to use Carbide formation For metal–metalloids (50 elements)	Serious spectral interferences Sub-ppm-ppb(μg/L) DLs* Expensive to run Also for some nonmetals (>70 elements)

AAS: atomic absorption spectrometry; *DLs: detection limits; ICP-OES: inductively coupled plasma-optical emission spectrometry.

comparative assessment, in the light of many years of development of the main pros and cons of each of these three popular and routine techniques.

Taking a risk inherent in excessive generalizations, perhaps we could say that FAAS dominates elemental inorganic analysis carried out in comparatively small laboratories and when only a few analytes (probably at mg/L, i.e., ppm levels) have to be determined.

When ppb level (μg/L) sensitivity is required, the technique of choice is ETAAS, at a cost of simplicity, robustness, and speediness of the analysis as compared to FAAS.

ICP-OES appears to have become the most popular routine technique for inorganic multi-elemental analysis of dissolved samples, even if initial investment and subsequent running expenses are much higher than those needed for AAS.

Of course, flow injection analysis (FIA) strategies have become commonplace in most laboratories opening new perspectives to sample handling, pretreatments, and manipulations for atomic spectrometry methods.

AAS instruments, flame and ETAAS based, are today widespread all over the world, both in developed and developing countries. They are less complex systems than multi-element techniques (e.g., ICP-OES) and so instrumentation for single-element AAS is considerably less costly. Thus, this book is devoted to those instruments and their applications.

1.5.2 Direct Solid Analysis Techniques

The rapid growth of ICP-OES might obscure the fact that the market size for direct solid analysis is still close to that of the dissolved samples. Spark source-optical emission spectrometry (SS-OES) and X-ray fluorescence (XRF) spectrometry are very well-established routine techniques that play the most important role in industrial control analytical chemistry of solid materials. Their importance today to monitor industrial processes, metallic and non-metallic raw materials, and final products, cannot be overemphasized; both techniques are widely implemented in routine application laboratories. More recently, other spectrochemical sources such as lasers and also glow discharges with optical emission spectroscopy measurements are also gaining momentum for the direct analysis of metallic, nonmetallic, and semiconductor solid samples.

While SS-OES is clearly an atomic technique, in which atoms are formed directly from the solid sample by virtue of the high energy of an electrical spark, the inclusion of XRF among the atomic techniques deserves some explanation. In this latter technique, the solid sample is interrogated about its composition at room temperature and so no free atoms are formed during the analysis. In XRF, a primary beam of X-rays is used to excite and eject electrons of inner shells of the atoms (e.g., shells K or L) of a solid. After such excitation, electrons of the outer shells "fall down" spontaneously to fill in the "holes" originated and the energy difference for such inner electron transitions is emitted as electromagnetic radiation (fluorescent or secondary X-rays). The information given by a fluorescent X-ray spectrum relates basically to the elements in the sample (not to the molecules). In other words, although strictly speaking XRF spectroscopy is a technique in which the analyzed sample is not transformed into single atoms, the information provided is mainly atomic (i.e., of the individual elements present in the solid sample as such).

Finally, to complete the picture of current atomic spectrometric techniques for inorganic elemental analysis, due reference to the increasing analytical importance of inductively coupled plasma-mass spectrometry (ICP-MS) should be made here. As suggested by the name, ICP-MS is the combination of an ICP with a mass spectrometer. During ICP-OES development it became clear that argon ICP was also a most efficient ion source to generate singly charged ions from the elements within a sample. It happened also that the kinetic energies of the ions formed in the argon plasma were between 2 to and 10 electronvolts, an energy interval most appropriate to be resolved by an inexpensive quadrupole mass analyzer. In this situation, the tremendous success achieved by coupling the argon ICP to such a widespread mass spectrometer is not surprising. The analytical performance characteristics of such a synergic combination, ICP-MS, are usually striking. As Table 1.3 shows, in a comparative manner, ICP-MS performance is clearly superior to its brother multi-elemental technique, ICP-OES. Apart from ICP-MS sensitivity and its relative spectra simplicity, probably its most differential (advantageous) feature versus photon-based measurement techniques is the ICP-MS capability to determine isotopes and to measure readily isotopic ratios. This capability gives to ICP-MS exclusive applications including the flourishing ones today of isotope dilution-ICP-MS methods for:

(a) Accurate determination of trace and ultratrace levels in the most and varied sample types.

(b) Stable isotope tracer use in living organisms in order to follow the metabolism of essential and toxic elements. Traditionally, radioactive tracers were employed for this purpose and most of the present knowledge about metal metabolism derives from experiments based on the administration of the corresponding radioactive isotopes to the studied living organism. Their eventual transport and distribution in the different organs is usually followed by radiochemical measurements. Of course, application of this radiochemical technique in humans has always been hindered by the detrimental effects of γ radiation to the body. With the advent of ICP-MS, a parallel strategy using nonradioactive isotopes seems straightforward.

Table 1.3. ICP-MS and ICP-OES main relative analytical merits

ICP-MS and ICP-OES (versus AAS)
• Multi-elemental character (simultaneous multi-element analysis possible).
• High speed of analysis.
• Semiquantitative rapid analysis (easier by ICP-MS).
• Continuous operation.
ICP-MS exclusive features
• Elemental detection in the ng/L range (or below).
• Simple mass spectra.
• Isotope measurement capability allowing for:
– Confirmation of the presence of multi-isotopic elements.
– Isotope dilution analysis.
– Stable isotope tracer applications (e.g., to follow the metabolism of elements).

AAS: atomic absorption spectrometry; ICP-MS: inductively coupled plasma-mass spectrometry; ICP-OES: inductively coupled plasma-optical emission spectrometry.

To conclude this section, however, it must be pointed out that ICP-MS is still a comparatively expensive analytical tool with a high price and maintenance costs, and its adequate operation demands highly qualified personnel. Both factors hinder its widespread use in many routine laboratories.

2

Theory and Basic Concepts in Atomic Absorption Spectrometry

The basic components to carry out atomic absorption spectrometry (AAS) measurements are shown and their roles discussed. A thorough description of the most common method to measure "absorption" in an atomic absorption process, the Walsh's method, is then given.

To ensure final accurate quantitative analysis by AAS, the types and nature of possible interferences in flame-AAS methods (spectral, physical, chemical, ionization, light scattering, and unspecific absorptions) are clarified. The chapter ends with a description of the analytical performance characteristics (sensitivity, detection limits, analytical working range, selectivity, accuracy and precision, cost, sample throughput, and availability of well-proven methodologies) of modern AAS.

2.1 GENERAL INTRODUCTION

We should start by stressing that atomic absorption spectrometry (AAS) has continued to be the most popularly utilized atomic spectrometric technique for routine elemental determinations for more than 40 years since commercial instrumentation became available. This fact indicates the existence of the practical strengths of AAS when compared with other atomic spectrometry techniques (see Table 1.2 in Chapter 1). After such a long period of evolution the theory and basic concepts of "analytical" AAS (as conceived and developed by Alan Walsh) are very well established. Thus, it is not surprising that there has been a clear slowdown in research undertaken with AAS, on the development of new methods, and on applications as well. The "turning point" in AAS publications seemed to have occurred at the end of the 1980s, a fact taken as the basis for a controversial and pessimistic prediction of AAS decline and "death" by the year 2000 (Hieftje 1989).

At the beginning of this new millennium, however, we may say that atomic absorption–based techniques are still "alive and kicking," particularly considering the high number of commercial AAS instruments used and still sold worldwide. New analytical AAS concepts and developments are still being investigated in some research laboratories (Evans et al. 2013). Among such efforts, the attempt to transform AAS into a viable and competitive multi-element method of analysis deserves a special mention. In fact, the use of continuum sources in combination with high-spectral-resolution systems (e.g., *echelle* spectrometers and charge-coupled devices [CCDs]) is now well established (Welz et al. 2010) and forms the basis of new commercial instrumentation available today in the market.

A second new concept and development (not yet commercialized) derives from the use of diode lasers, instead of classical hollow cathode lamps, to reduce the size and cost of future AAS instruments (Zybin et al. 2005). Another field of rather basic research is the work on new concepts and instruments arising from the use of conventional lasers in connection with cavities (as atomizers) to enhance the absorption and hence the sensitivity of possible AAS determinations (Emig et al. 2002).

Studies on introducing volatilized analytes to the atomizer, or on graphite furnaces as alternative atomizers, are still appearing in the literature, while renewed efforts are being devoted to modernization, automation (e.g., with flow techniques), and also, to miniaturization of conventional AAS instruments in several laboratories around the world.

In spite of these still-standing research activities, it is fair to say that the real strength of AAS today—its proven reliability, robustness, and efficiency for real-life elemental analysis—derives mainly from the original Alan Walsh invention (his first patent in AAS was granted in 1953) in which the concept of using a hollow cathode lamp as the light source for absorption measurements was decisive to the success of the technique. This exceptional light source was proposed originally in combination with a flame, as the atomizer of the sample to be analyzed. Today it is clear that the nature of that light source and of the atomizer employed will determine the actual strengths and shortcomings of the AAS analytical methods based on such instrumentation.

To understand why the hollow cathode lamp performance is central to the success of AAS elemental determinations, why the type of atomizer used will determine dramatically the detection limits (DLs) achieved, or why the

Source   Atom cell   Monochromator   Transducer & Readout

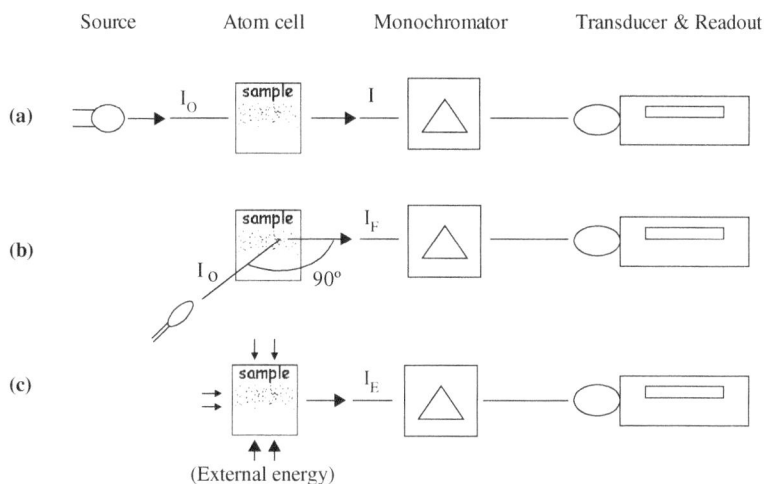

Figure 2.1 Schematics of the basic atomic absorption spectrometry experiment (a) and the main differences with the other atomic spectrometry main techniques of fluorescence (b) and emission (c).

necessary straight-line geometry of AAS optics makes this technique typically usable for only one (or just a few) element(s) at a time (perhaps its main drawback when compared with inductively coupled plasma-optical emission spectrometry [ICP-OES]) are of utmost importance for an intelligent and efficient use of AAS in the laboratory. To help to understand such practical issues, which are of critical importance to this book, the fundamental concepts of AAS measurements and its inherent analytical strengths will be considered in the following sections.

2.2 THE BASIC ATOMIC ABSORPTION SPECTROMETRY EXPERIMENT

As explained in the first chapter, atomic absorption is the process that occurs when a ground-state electron of the desired atom (analyte) absorbs energy, in the form of light of a given frequency, from an external light source and so the electron is excited to an upper-energy-level state of the atom. Thus, the basic AAS experiment is easily illustrated by the simple schematics shown in Figure 2.1a. Such a self-explanatory schematic, common to all absorption spectroscopic techniques, shows that the main components needed to carry out AAS measurements are:

1. The light source (which provides the external energy).
2. The atom cell (where the sample is atomized).

3. The monochromator (where the appropriate frequency is selected).
4. The "light-read-out" system, integrated by a transducer or detector (trans-forming the light into a measurable signal, e.g., an electric current) and the electronic read-out system of such measured signal.

With such a system adequately assembled (see Figure 2.1a) the amount of light energy absorbed by the analyte atoms at the selected wavelength can be easily measured by undertaking two consecutive measurements: the incident light intensity or blank (I_0), without the sample in the atom cell, is measured first, and then the light intensity exiting the atom cell (I), when the sample (actually the analyte atoms in the light path) absorbs such light.

In the absence of spurious phenomena (interferences) the amount of light energy absorbed by the analyte present in the sample will be $I_0 - I$. Such an amount, $I_0 - I$, will increase as the number of analyte atoms in the light path increases. In fact, as we will see later, there is a relationship between the measured absorption and the analyte concentration in the original sample. Once the exact relationship between those two magnitudes is clearly established (e.g., using known analyte concentrations of adequate standards), that is, once the calibration curve is defined, analyte concentrations in unknown samples can be determined by measuring the amount of light they absorb. Figures 2.1b and 2.1c illustrate, comparatively with the AAS experiment, the differences among the three main atomic spectrometry techniques: Figure 2.1b shows that in atomic fluorescence spectrometry (AFS) the basic instrument components are the same, only the geometry of the AAS arrangement changes as the external light source (used for photoexcitation of the analyte) is rotated 90° with respect to the straight-line optical axis used in absorption measurements. The AFS measurement then consists of measuring the fluorescence emitted at a given solid angle (I_F) by the excited atoms.

Finally, as shown in Figure 2.1c, the atomic emission basic experiment is the simplest one because it does not need an external light source; the excited sample in the atomizer acts as such. Here, the basic components, except the source, are again the same. However, now, some type of energy device (provided by a spark, a flame, a plasma, etc.) is needed to excite the sample for atomic emission spectrometry (AES, also known as OES from "optical emission spectrometry"). Then, the spontaneous emission radiation from excited analyte atoms occurs, allowing for the amount of light emitted by them (I_E) at the selected frequency to be finally measured.

Of course, as for the absorption process, both I_F and I_E can be related to the analyte concentration in the sample and so the corresponding "calibration curves" can be established. As mentioned for AAS measurements, both alternative experimental arrangements (b and c) could eventually allow for determining the analyte concentrations in unknown samples. In practice, they constitute the basis of the other two main atomic measurement arrangements, namely atomic fluorescence and atomic emission.

2.3 THE ABSORPTION COEFFICIENT CONCEPT

There are different methods of measuring absorption in an atomic absorption process, depending on the experimental magnitude measured:

1. The "integrated absorption," where the area under the atomic absorption line profile (see Figure 1.6) is measured.
2. The "total energy absorbed" from a continuum source, emitting a continuum spectrum, at the absorption line.
3. The "line absorption" method, by which we measure the relative absorption of light, at the absorption wavelength, from a source that emits a line spectrum.

The latter method is the most practical one, called the Walsh method, named after its inventor. This measurement method constitutes the basis of routine AAS measurements today for analytical determinations. In such analytical AAS measurements, a line source (usually a hollow cathode lamp) is used as the critical component of the AAS experiment. As indicated in the preceding chapter, atomic absorption spectra, as opposed to the classical absorption bands observed in conventional molecular absorption spectrophotometry, show up in the form of several individual narrow lines. The frequency (wavelength) of such lines characterizes the particular absorbing atoms (analyte). If a hollow cathode lamp of the particular element to be determined is used, a line spectrum of the analyte is obtained from which an adequate emission line is selected with a monochromator (wavelength maximum, λ_0; intensity, I_0).

This light is passed through the atomic cloud of free atoms of the analyte (in the flame), contained in an absorption volume of path length L. The atomic vapor in the flame absorbs the radiation emitted by the hollow cathode lamp virtually at the maximum of the element's absorption line, λ_0. As shown in Figure 2.1a, the intensity emerging from the atomizer, I (that part of I_0 which was not absorbed), is called the "transmitted" light and follows the general law of absorption, the so-called *Beer's law*:

$$I = I_0\, e^{-k_\upsilon\, l}$$

where k_υ is defined as the absorption coefficient at the wavelength of measurement of the atomic vapor considered and l the optical path in the *cuvette*. The magnitude of k_υ characterizes absorption lines as strong, medium, or weak and, for a given element, will depend on the frequency of the selected line (analytical line) and on the concentration of the analyte absorbing atoms. For general instrumental quantitative analysis we usually measure a property of the desired atoms (or molecules) that varies linearly with the concentration of such atoms (species) of interest in order to establish a typical calibration curve.

As Beer's law shows, it depends exponentially on k_υ. For this reason, it is commonplace to use a more convenient form of this formula showing a linear relationship between the measured property and the desired values of analyte concentration. Such a property is the absorbance, Abs, defined as:

$$Abs = -\log(I/I_0) = \log I_0/I$$

Thus, if this concept of Abs is introduced in the previous Beer's Law equation we get:

$$Abs = \log I_0/I = \log e^{k_\upsilon l} = 0.4343\ k_\upsilon l \tag{2.1}$$

Therefore, *Abs* is linearly related to k_υ (i.e., at a given λ_0, with the atomic concentration of the analyte in the atomizer) and to the optical path of the light through the analyte atoms, l.

If the analytical or line method of absorbance is used, as proposed by Walsh when using a hollow cathode lamp, the absorbance measurement is very simple and practical: The I_0 measurement needed (see equations shown earlier) is carried out virtually at the maximum, λ_0, of the absorption atomic line because the emission line from the hollow cathode lamp at the λ_0 will be much less broadened than the corresponding analyte absorption line in the flame (see Note below).

Specifically, the emission line halfwidth of the lamp is much smaller than the halfwidth of the corresponding absorption line in the flame, at the same wavelength. This fact, characteristic of the Walsh method, is illustrated in Figure 2.2, which shows how absorption coefficients measured will correspond to those at the maximum of the absorption line profile, K_0^ν. Therefore, the K_0^ν value could be evaluated from the Doppler theory, which explains well the broadening in the center of atomic lines observed in flames (as explained in Chapter 1).

If the value of Abs is so measured at λ_0 maximum (line method), the random thermal motion of analyte atoms relative to the observer is governed by a Maxwell distribution in the thermal equilibrium. Under such conditions, the Doppler effect demonstrates that K_υ distribution with wavelength can be expressed by a formula that at its maximum value, K_0^D, is given by the expression:

$$K_0^D = \frac{2\lambda_0^2}{\Delta\lambda_D} \cdot \sqrt{\frac{\ln 2}{\pi}} \cdot \frac{\pi e^2}{mc^2} \cdot f \cdot N_0$$

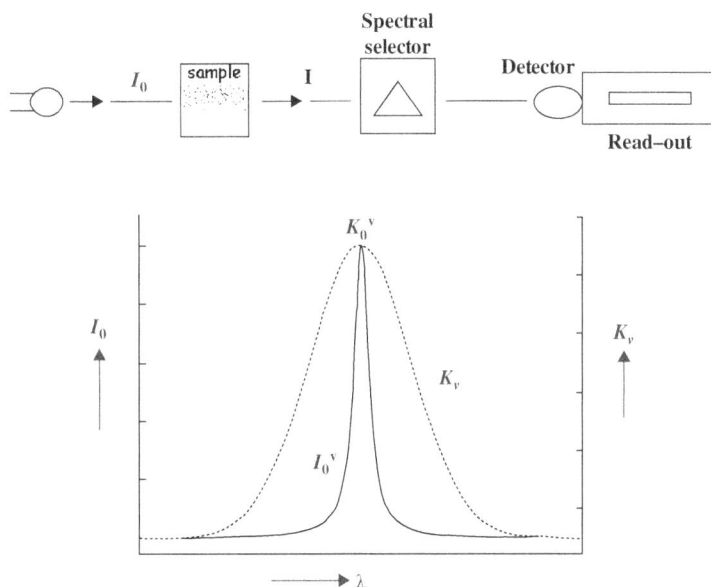

Figure 2.2 Basic atomic absorption spectrometry measurements and the "lock-and-key" effect using a hollow cathode lamp.

While a hollow cathode lamp operates at low pressures and temperatures, and hence emission profiles, I_E, will not be too much broadened, a flame burns at atmospheric pressure (760 torr) and at 2,000–2,500°K temperatures. In other words, the Lorentz and Doppler effects will be much higher in a flame and K_υ more broadened than in the hollow cathode lamp atomic clouds (see Figure 2.2 for visual illustration).

where:
$\Delta\lambda_D$ = Doppler halfwidth of the absorption line used
e = charge of the electron
m = mass of the electron
c = light speed in the vacuum
f = oscillator strength of the absorption line used
N_0 = analyte atom density (number of atoms per unit volume) in the ground state in the atomizer.

Now, if we introduce this value of K_0^D in the general expression of Abs we get:

$$\text{Abs} = 0.4343 \frac{2\lambda_0^2}{\Delta\lambda_D} \cdot \sqrt{\frac{\ln 2}{\pi}} \cdot \frac{\pi e^2}{mc^2} \cdot f \cdot l \cdot N_0$$

Thus, for a given set of experimental conditions in an absorption experiment, where the magnitudes of most of the terms in the above equation are constant, we get:

$$\text{Abs} = \text{constant} \cdot l \cdot N_0 \qquad (2.2)$$

and, at constant path length or optical path l, absorbance is directly proportional to the analyte concentration in the atomizer.

2.4 QUANTITATIVE ANALYSIS BY ATOMIC ABSORPTION SPECTROMETRY

Equation (2.2), derived earlier for the analytical absorption of atoms in a flame, resembles very much the general Beer's law expression, considered as the theoretical basis of molecular absorption spectrophotometry, that is:

$$\text{Abs} = \log I_0/I = \varepsilon \cdot l \cdot c \tag{2.3}$$

where "ε" is a dimensionless constant called *molar absorptivity* (or absorption coefficient), "l" is the optical path length through the absorbing sample (in cm), and "c" is the analyte concentration (mol/L).

The analogy of Equations (2.2) and (2.3) is apparent except for the way in which the analyte concentration is given: N_0 relates to that concentration in the atomizer (in the gas phase of the flame), while "c" refers to actual concentration of the absorbing compound (analyte) in the sample solution. Therefore, Formula (2.2) may become Formula (2.3), which is the mathematical expression of the fundamental law for absorption spectrophotometry, if the dependence between N_0 and "c" is clearly established.

Such a relationship between the atom concentration per unit volume in the flame ($N_T \approx N_0$) and the analytical concentration of the sought analyte in the sample solution, "c" (which is aspirated and nebulized from the solution into the flame; see Figure 2.3), can be approximated by an empirical formula of the type:

$$N_0 \cong N_T = \text{constant} \cdot \frac{F \cdot \varepsilon \cdot \beta}{Q \cdot T} \cdot c \tag{2.4}$$

where the different experimental variables (whose values will be determined by the nebulization of the sample into the flame) determining the final atom population are as follows:

F = flow of the aspirated sample solution (mL/min)
ε and β = efficiencies of vaporization and atomization processes in the flame, respectively
Q = flow of the gas mixture in the flame (before burning, mL/s)
T = absolute temperature in the atomizer (°K).

Therefore, working in the laboratory under fixed nebulization conditions for a given atomic line of the analyte, it follows that we may say that N_0 is

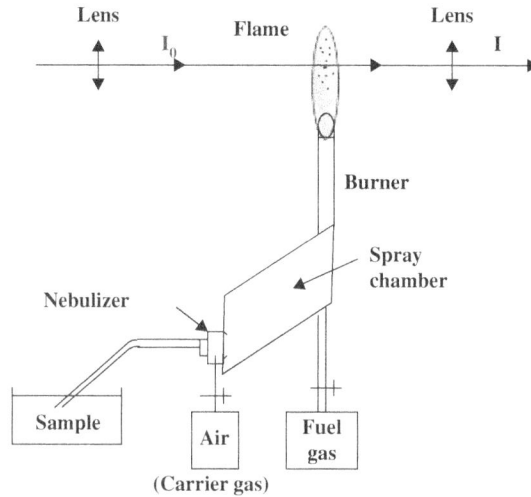

Figure 2.3 Classical sample introduction system for flame atomic absorption spectrometry measurements (*via* nebulization of the dissolved sample).

directly proportional to "c" in the solution ($N_0 \sim k'c$). In other words, if we combine Equations (2.4) and (2.3), we obtain the formula:

$$\text{Abs} = k \cdot k' \cdot l \cdot c = \text{constant} \cdot l \cdot c \qquad (2.5)$$

This basic equation of AAS is a particular case of the general Beer's law of light absorption. In brief, from a practical point of view, by working in the laboratory under fixed experimental conditions, including optical path, l (as it is customary in the practice of real AAS analysis), the measured analytical signal of Abs is directly proportional to the analytical concentration of the sought element in the sample solution. This analytical signal depends also upon other parameters (i.e., important experimental variables) that we have discussed to arrive at the Formula (2.5) given earlier. For practical work in the laboratory all those experimental variables should be known, optimized for each analyte determination, and eventually fixed accordingly. Then, the calibration curve, which is the linear plot of "Abs" versus "c," can be determined.

As such a dependence may also be so complex, in practice, we define the function of dependence experimentally for the fixed conditions of the analysis measuring the Abs values observed for several standards of increasing and known concentrations of the analyte. Once the calibration curve (see Figure 2.4, plot a) has been obtained experimentally, we may now obtain the values of

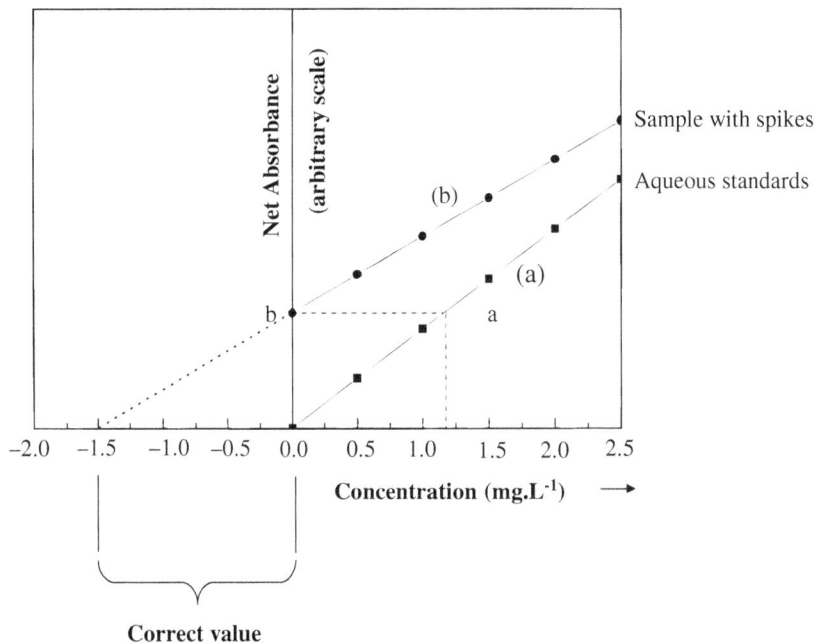

Figure 2.4 Main calibration modes in atomic absorption spectrometry. (a) Conventional aqueous calibration. (b) Standard addition calibration to correct for matrix interferences.

"c" in any unknown sample solution, just by resorting to such a calibration. Such direct calibration is enough for many applications, but it is not valid in the presence of matrix effects (see the multiplicative interferences concept and how to tackle them in Box 2.1).

Once the blank values are subtracted from the total Abs measured, the linear relationship is given by: Abs = $k \cdot c$ and this is shown in the plot termed aqueous standards of Figure 2.4a.

Sometimes, however, the matrix of a given sample may modify the slope of this straight line obtained with pure aqueous standards. In such cases, direct calibration would introduce important analytical errors (due to the so-called multiplicative interferences) unless exact matrix matching is used for the calibration.

Exact matrix matching is not always feasible as the precise matrix composition may be unknown or not completely known. In such cases, the so-called *standard additions* method can be used: The sample is spiked with a few additions, five in the case shown in Figure 2.4 plot b, of known amounts of analyte in such a way that the sample matrix is not significantly changed. Then the absorbance of the unspiked and the spiked, adequately diluted, samples is measured. As shown in Figure 2.4 plot b, by extrapolation back to the negative side of abscissa, the unknown and correct value of the analyte in the sample may be calculated (1.5 mg/L in the figure rather than 1.1 mg/L, as obtained using the Abs = $k \cdot c$ plot obtained for "aqueous calibration").

Of course, any slope differences (comparing the slope observed for standards with analyte spikes added to the sample) is a decisive proof of matrix or multiplicative interferences, hence the need for using standard additions for accurate determinations.

2.5 INTERFERENCES IN FLAME ANALYTICAL ATOMIC SPECTROMETRY TECHNIQUES

In order to understand properly the exceptional analytical selectivity characterizing flame-AAS methods, a basic knowledge of the different sources of interferences that may be encountered in atomic spectrometry is a must.

Reviewing the main interferences occurring using a flame as atomizer (in the three different measurement modes schematically shown in Figure 2.1) may reveal the general sources of error to expect in all known atomic spectrometry measurement techniques. The main types of such sources of error are summarized in Box 2.1 for absorption, emission, and fluorescence modes.

Box 2.1 Main types of interferences (possible sources of error) in atomic spectrometry

- Affecting the three techniques (AAS, AES, AFS):
  - Spectral
  - Physical
  - Chemical
  - Ionization
  - Variations of the temperature of the atomizer
- Affecting absorption and fluorescence modes:
  - Scattering of incident light
- Affecting fluorescence (AFS) mode only:
  - Fluorescence quenching

The concept and relative magnitude for each type of interference will be described next and compared for the three main atomic modes using the flame as the atomizer of reference. The following discussion is a basic platform to understand and assess potential sources of error in any atomic technique we might need in our laboratory (i.e., concepts can easily be extrapolated to using an atomizer different from the flame).

2.5.1 Spectral Interferences

These are interferences brought about by the detection of a radiation that is due to or coming from elements (or molecules) different from the sought analyte. In other words, such "interfering" radiation spectrally overlaps the measured atomic line of the analyte within the used spectral bandwidth of the used monochromator. Of course, a positive analytical error would be observed should such an interference occur.

2.5.2 Physical (Transport) Interferences

This source of interference is particularly important in flame-based methods because it derives from the need (see Figure 2.3) that the liquid sample must be aspirated and transported reproducibly, as a fine aerosol, into the flame.

Changes in the solvent, viscosity, density, or surface tension in the sample solution will affect the final efficiency of nebulization and transport processes. Therefore, any of these changes or a deficient nebulization, changing the effective transport of the analyte, as compared to nebulization and transport obtained initially for the standards used for calibration, will modify the final density of analyte atoms in the flame, even if their concentration in the sample is the same.

2.5.3 Chemical Interferences

Using a comparatively low-temperature atomizer (as flames or graphite furnaces are compared to an ICP) these types of interferences are probably the most serious ones. Their possible occurrence must be particularly assessed for accurate determinations.

Chemical interferences in atomic spectrometry are due to the presence or formation in the atomizer of analyte refractory compounds. Of course, when the temperature in the atomizer is not high enough to dissociate and atomize completely the analyte, a loss of the expected analyte atom population will occur. The corresponding reduction of analyte atoms (those trapped in the refractory molecule) will bring about a significant decrease of the atomic analytical signal.

Typical examples are phosphate interferences in blood serum determinations of Ca and Mg by flame-based atomic spectrometry methods. Phosphates of these metals can form and they are only partially dissociated and atomized at normal flame temperatures.

Another good illustration of the nature and importance of chemical interferences in flames is, for instance, that elements such as Al, Si, Zr, Ta, Nb were thought, in the first years of AAS development, to be nonaccessible to AAS determinations. For several years, no atomic signals were obtained for such "refractory" elements in a common air/acetylene flame. In practice, only the introduction of "hotter" flames constituted by nitrous oxide/acetylene explained correctly the reasons for that, solving the problem of required dissociation of the corresponding refractory oxides and hydroxides to form analyte atoms:

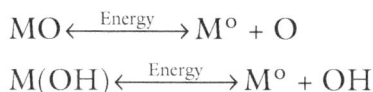

$$MO \xleftrightarrow{\text{Energy}} M^{o} + O$$

$$M(OH) \xleftrightarrow{\text{Energy}} M^{o} + OH$$

The above dissociations could not be achieved at around $2{,}000^{o}K$ with a conventional flame, but the higher temperatures occurring inside a N_2O/C_2H_2 flame $(2{,}800–3{,}200^{o}K)$ could dissociate the refractory oxides and hydroxides to provide free M^{o} atoms, enabling strong, analytically useful, atomic signals for the determination of such elements.

Apart from using "hotter" flames and atomizers, an alternative and efficient way to overcome chemical interferences is to resort to "releasing agents." These are chemical reagents (e.g., organic chelating compounds, such as 8-hydroxyquinoline) able to form compounds with the analyte (competing with the refractory ones), which are easily dissociated at the usual temperatures of an analytical flame.

2.5.4 Ionization Interferences

Particularly when using *hot* flames, such as the N_2O/C_2H_2, an opposite phenomenon may occur, which will, however, also reduce the free atoms' population: if a low ionization energy metal, M, is introduced into a hot atomizer, complete loss of the electron from the neutral atom is possible:

$$M^{o} \xleftrightarrow{\text{+Energy}} M^{+} + e^{-}$$

in such a way that ionization may take place and the M^{o} population will decrease (i.e., the sensitivity of the analyte determination, for which an atomic line of M^{o} is used, is proportionally reduced).

Fortunately, these ionization interferences can be suppressed by adding to the sample solution another element or compound that can provide a great excess of electrons in the flame (i.e., another easily ionizable element). In this way, the above ionization equilibrium is forced to the left (of the equation).

Well-known examples of the use of such buffering compounds are salts of Cs and of La (easily ionizable elements), which are widely used as ionization buffers in the determination of metals such as Na, K, or Ca by flame-AAS (or flame-OES).

2.5.5 Temperature Variations in the Atomizer

Variations in the temperature of the atomizer would change the population of neutral atoms (M^o) needed for atomic absorption, but particularly of excited atoms (M^*), essential for atomic emission measurements. Therefore, the use of atomizers in which a constant temperature is maintained during operation should be aimed for in order to ensure reproducible analytical signals.

2.5.6 Light Scattering and Unspecific Absorptions

Both types of problems are only observed when an external line source is used (i.e., as Figure 2.1 shows, just in AAS and AFS measurements). When part of the light coming from the hollow cathode lamp, I_0, is scattered by small particles in the flame (e.g., droplets or refractory solid particles) or perhaps absorbed unspecifically (e.g., by undissociated molecules existing in the flame) important analytical errors would be derived if no adequate corrections are made of such unwanted occurrences: first, because the scattered or dispersed radiation diminishes the real hollow cathode lamp intensity, I_0, thereby affecting the measurement. More importantly, false analytical signals could be measured and so confused with the specific absorption due to the analyte atoms. If dispersed radiation or unspecific molecular absorptions are detected, those signals are not due to the analyte and would result in gross errors.

Fortunately, both sources of false signals can be easily distinguished from the "specific" analyte signals, which occur only at the analytical line. This basic differential feature can be adequately used for background correction.

2.5.7 Quenching of the Fluorescence

For the sake of completeness it is appropriate to finish this section by referring to a type of interference that is only of interest in fluorescence type of measurements (see Figure 2.1b). Quenching is a deactivation of excited states in atoms (or molecules) due to collisions with the surrounding atoms or species existing in the atomizer (e.g., coming from the burning flame or the matrix of analyzed samples). This deactivation is nonradiational and hence it will decrease the intensity of the fluorescence observed.

This fluorescence quenching can be minimized by using an inert gas environment for the analyte atoms. If an inert noble gas surrounds excited atoms, quenching is very small as compared with molecular gases. Thus, flames "purged with Ar" have been recommended for flame atomic fluorescence measurements.

2.6 ANALYTICAL PERFORMANCE CHARACTERISTICS OF AAS

As pointed out already, the most popularly used and sold experimental arrangement of the three possible measurement modes illustrated in Figure 2.1 is, still today, the one featuring those components and arrangement characterizing an atomic absorption spectrometer. In consideration of and with the practical perspective offered by more than half a century of routine work with those flame-based atomic techniques, this has to be linked to the problem-solving capability that each of the three arrangements (allowing for absorption, fluorescence, or emission measurements) offers to the analytical chemist. In other words, in the long term, the eventual success of one or more of such three techniques should be determined by the intrinsic analytical performance characteristics of each measurement mode.

It is true that the three of them are in a way complementary for application to solve problems requiring elemental analysis information. As explained in Section 1.5 of the first chapter "there are horses for courses" (i.e., we will need different techniques for different analytical problems) and so we should not buy atomic techniques for dissolved-sample analysis if the main task of our laboratory is direct solid analysis.

Thus, it may not always be clear which atomic technique is optimum to tackle all the analytical problems faced within a routine analytical laboratory. However, it is perfectly clear that to decide on the best-suited one for your particular laboratory you need double knowledge: first, the real problems and elemental analysis type to be faced must be well defined, and second, once the analytical requirements in your laboratory have been clearly established, knowledge of the analytical performance and capabilities of the different available atomic techniques is mandatory for an intelligent and informed selection.

In order to compare analytical techniques of different types, the use of some critical parameters, worldwide accepted as comparison key criteria, is now commonplace. In this context, sensitivity, DLs, analytical working range, selectivity, accuracy and precision, cost, sample throughput, and availability of well-proven methodologies are criteria whose concept and relative importance must be clear to the eventual user of those techniques. Using the same atomizer (e.g., a flame, which is still the most popular one) the techniques derived from the three basic designs depicted in Figure 2.1 exhibit fundamental advantages and limitations that should be properly understood.

Therefore, those comparative key criteria and their characteristics for the three flame techniques will now be compared, as such a discussion is of fundamental importance, not only to understand flame techniques but to introduce fundamental concepts (which can be extrapolated to any instrumental analytical technique) for eventual analytical performance characteristics and method validation and comparisons.

2.6.1 Sensitivity and Detection Limits

Analytical sensitivity, S, refers to the slope of the linear calibration graph in an instrumental method of analysis. In AAS methods, sometimes the term "sensitivity" means "the concentration of the analyte producing an Abs = 0.0044 (1% absorption)," and is a form of assessing the value of S.

To clearly define the detection power of a given technique or method, which is the lowest detectable concentration of the analyte with such a technique, a more general concept is used: that is, the DL of the technique or method for a particular element. The DL informs one of the minimum concentrations (or amount) of the analyte needed to be detected, with a given level of certainty, by the considered method. Based on statistical considerations (taking into account the degree of certainty required), it has become customary to define the DL following the IUPAC criterion: The DL of any analytical method refers to "the concentration (amount) of the analyte producing a net signal three times greater than the value of the observed standard deviation of the blank (i.e., the sample without analyte) signal, σ_B. In other words, C_{DL} means the minimum concentration value of the analyte in the sample that produces a signal distinctly higher, with a known probability, than that observed for the blank.

At the C_{DL} level and in Abs measurements:

$$\text{Abs}_T^{DL} = \text{Abs}_{blank}^{DL} + 3\sigma_B \ \text{ or } \ \text{Abs}_{net}^{DL} = 3\sigma_B$$

As the sensitivity is given by the slope of the calibration graph, $S = \dfrac{\text{Abs}_{net}}{C}$, at any given analyte concentration, we have:

$$S = \frac{\text{Abs}_{net}}{C} = \frac{\text{Abs}_{net}^{DL}}{C_{DL}}$$

A good estimation of the "detection limit" (C_{DL}) is:

$$C_{DL} = \frac{\text{Abs}_{net}^{DL}}{\text{Abs}_{net}/C} = \frac{3\sigma_B}{S} \tag{2.6}$$

This formula shows that C_{DL} is better (lower) not only when S of the technique is higher but also when the standard deviation of the blank, σ_B, is smaller.

Sometimes the term "determination limit" or limit of quantification (L_Q) is also used to express the minimum concentration needed for a reliable quantitative determination of the analyte. Again, based on statistical considerations on the acceptable uncertainty of the value given, the valuable parameter is defined by IUPAC as $L_Q = 10\dfrac{\sigma_B}{S}$.

Probably, the C_{DL} values are the most common approach used to compare several analytical techniques (or methods) from the point of view of their detection power for a given analyte. Unfortunately, C_{DL} values in the literature are often misleading (usually too optimistic). Perhaps, for practical purposes in the laboratory, L_Q values may provide a more realistic figure in methods' evaluations.

It is important to have a relatively complete comparative picture of the typical DL ranges for the most important atomic spectrometry techniques available today for routine elemental analysis. Figure 2.5 provides such a picture for the three AAS techniques discussed in this book, as compared to more sophisticated ICP-based atomic techniques. It is clear that the best DLs are provided by ICP-MS but graphite furnace-AAS and hydride generation-AAS may offer in some cases extremely good DLs at a comparatively low cost.

Among flame-based atomic techniques, flame-AAS proves to be the most versatile technique as it is applicable to the widest range of elements over a wide range of concentrations. Flame-AES can be adequate for alkali and alkaline-earth metals, while flame-AFS might provide the best sensitivity for others (including metals of specific clinical interest including Hg, Cd, Pb, or Tl).

2.6.2 Selectivity of the Three Flame-Based Techniques

As pointed out earlier, it is useful to discuss the comparative selectivity of each technique by careful examination of the extent of the different types of

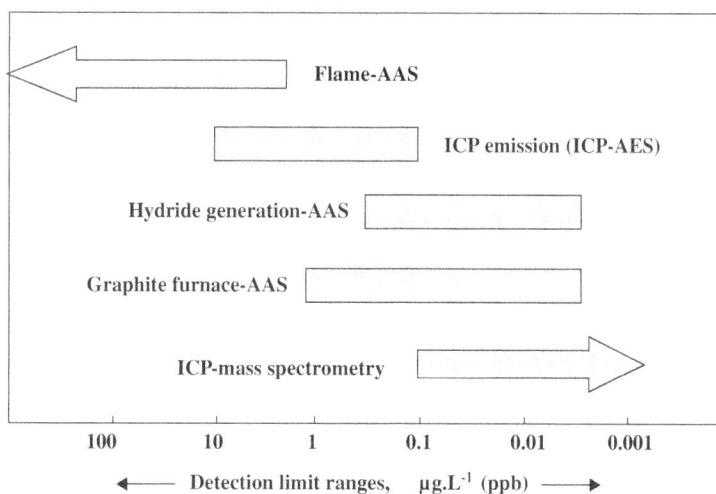

Figure 2.5 Typical detection limit ranges for the most important atomic spectrometry techniques.

interferences expected in each technique. A brief review of the relative magnitude of those interferences in each case is given below.

Considering *spectral interferences* it is fair to start stressing that atomic spectrometry provides much better spectral selectivity than the corresponding UV-VIS molecular absorption spectrophotometry methods (the halfwidth of atomic lines is orders of magnitude smaller and so spectral overlapping will be much less likely).

When the magnitudes of these interferences are compared for emission, absorption, and fluorescence, the lock-and-key AAS arrangement provides the most selective mode (working with a hollow cathode lamp, real line spectral interferences are virtually nonexistent even when using a relatively low-resolution monochromator). Atomic emission (Figure 2.1c) arrangements should be more prone to spectral overlapping (in practice, this is particularly true if the temperature of the excitation source is high, e.g., in ICP-OES, and many energy levels can be excited). Finally, regarding flame-AFS it can be said that using a hollow cathode lamp for excitation, specific spectral interferences are scarce in atomic fluorescence because only the emission lines of the analyte should bring about its atoms' excitation in the flame. In any case, when resonance fluorescence is used for the analysis (the same frequency is used for excitation and for the final fluorescence measurement) light scattering of I_0 in the flame would produce a most problematic spectral interference at the AFS measurement frequency. Of course, such an emission is unspecific and could be corrected for by using background correctors (based on measuring the unspecific background, away from the fluorescence atomic line, as we will see later on for AAS measurements).

Physical interferences, affecting the efficiency of the analyte transport to the flame, will usually decrease the observed AAS signal. Although less analyte is usually transported (e.g., because samples present higher viscosity than standards used for calibration), the reverse effect is also possible. If the formation of volatile compounds takes place during nebulization (e.g., for osmium, the formation of OsO_2 in the presence of HNO_3) AAS enhanced signals may be also observed. In this case, absorption, emission, or fluorescence modes will be equally affected by such alterations.

Chemical interferences are known to be most serious in a flame due to its comparatively low temperature. In any case, as the three modes depicted in Figure 2.1 need an efficient atom formation, the expected analytical sensitivity deterioration, due to refractory compound formation in the flame, will be identical for the three measurement modes.

The case of *ionization interferences* is analogous: if the measured atoms, M^o, disappear to form ions, M^+, the analytical signal will decrease but the observed magnitude of the decrease will be again identical for the three flame-based techniques schematized in Figure 2.1 (i.e., all of them rely on efficient and reproducible analyte atom formation).

Variations in the atomizer temperature will, of course, affect the AAS measurements. However, such variations will have an immediate effect on the excited-atom population, N_2 (i.e., we know that: $\frac{N_2}{N_1} \cong e^{-\frac{h\nu}{kT}}$, that is, the magnitude of N_2 varies exponentially with the absolute temperature in the flame). Therefore, maintaining a constant T during measurements is critical in flame-atomic emission methods.

For AAS measurements, the important parameter is N_0 (atoms' population at the fundamental level) and N_0 is barely influenced by small variations (e.g., $\leq 2\%$) in the flame temperature. In brief, errors derived from fluctuation in the flame temperature are less important in AAS than in AES measurements.

Light scattering interferences do not exist in emission, but they may be a source of analytical errors in AAS and AFS because we have to use an external light source of excitation (the lamp). Fortunately, all Rayleigh light scattering interferences are unspecific in nature (i.e., they occur at the atomic analytical line but also close to and away from that measurement line). Such problems are routinely tackled in an automatic manner using an adequate background corrector system, which is sold today as an accessory of modern AAS spectrometers. This constitutes a similar problem in AFS so adequate instrumentation should be used to correct any source of analytical errors.

2.6.3 Accuracy and Precision

It is known that the accuracy of analytical methods relates to the observed closeness between the measured value (the mean of a series of replicates, obtained by repeating the same analytical determination a known number of times) and the true value of the sought concentration (or amount) of the analyte. The difference between the two values gives us the analytical error. For AAS, the main sources of such error are detailed in Section 2.5. That is, the presence of interferences will bring about a bias of the measured value in a given direction (i.e., a positive interference will originate always positive errors).

However, even if an experienced operator repeats the same determination many times some variability of the obtained results is observed. This variability depends on many uncontrolled factors (it is random, and so unavoidable to a certain extent) but it can be addressed by the repeatability "standard deviation," SD, observed in the "n" repeats carried out, $\mathrm{SD} = \sqrt{\dfrac{\sum (x_i - \bar{x})^2}{n - 1}}$

where x_i is any individual measurement and $\bar{x}$ the mean value for all the repeats carried out.

The relative standard deviation is then $SD_r = \dfrac{\sigma}{\bar{x}}$; it is usually expressed as a % and termed "coefficient of variation," $SD_r(\%) = \dfrac{\sigma}{\bar{x}} \cdot 100$, of the mean.

Of course, an adequate control of interferences is mandatory in order to avoid "bias" in AAS determinations. The repeatability of our AAS measurements should be also optimized and eventually worked out.

Typical concentration levels determined by flame-AAS methods are in the range of a few mg/L and repeatability SD_r (%) values observed for such concentrations are always better than ±1%. This value is clearly lower (better) than the corresponding SD_r (%) observed using AES or AFS measurements.

2.6.4 Analytical Linear Range

The linear relationship between measured Abs and sought analyte concentration, c, is governed by Beer's law (Equation [2.5]), but the range of c values among which such relationship holds is not too large. There are deviations from the linear behavior due to several factors, particularly when c values are too high (typically three orders of magnitude above the corresponding limits of quantification). This particular concern is less troublesome for AES and AFS methods for which five or six orders of magnitude of analytical linear ranges are commonplace.

In brief, the analytical working range can be viewed as the concentration range of the analyte in the sample over which reliable quantitative results can be obtained without further recalibrations. As discussed before, AAS offers, generally speaking, more limited ranges than emission-based methods.

2.6.5 Versatility and Sample Throughput

About 70 elements of the Periodic Table can be determined at 0.1– to 100– ppm levels using modern flame AAS methods. However, AAS techniques are basically single-element techniques (particularly so for graphite furnace-AAS for which the lamp and the atomization conditions have, as a rule, to be selected individually for each element). This feature determines that sample throughput in AAS (especially with electrothermal atomization, as detailed in Chapter 5) is comparatively lower.

In practice, atomic emission techniques (particularly those using a hot spectrochemical source as in ICP-OES) are intrinsically multi-elemental and this characteristic offers the possibility of a very high sample throughput in routine analysis. To counterbalance such a clear advantage, however, AAS techniques are simpler and so considerably less costly that ICP-OES or ICP-MS techniques (which do offer high-throughput multi-elemental analysis).

2.6.7 Robustness and Availability of Well-Proven Methodologies

To conclude this chapter it is fair to say that flame-AAS is probably the easier-to-use and more robust of the three atomic techniques schematically shown in Figure 2.1. Extensive information on analytical methodologies and protocols for elemental determinations by flame-AAS is available from the AAS instruments' manufacturers (see Appendix A for details on 10 of such companies).

The AAS arrangement of Figure 2.1a provides a comparatively more robust measurement because the analytical signal measured is Abs = -log I/I_0. Abs is a ratio of two measurements of intensity, rather than an absolute emission intensity measurement, typically required for emission and fluorescence techniques. This ratio will cancel out many sources of variability affecting a single measurement.

In brief, AAS techniques may provide most robust analytical procedures to the point of representing today the more reliable atomic spectrometry measurement technique for elemental determinations at the appropriate concentration level (see Figure 2.5 for an overall comparison of such concentration ranges).

3

Basic Components of Atomic Absorption Spectrometric Instruments

Radiation sources, atomizers, monochromators, and detectors used in modern instruments are described. Common radiation sources emitting a narrow-line spectrum from the element of interest are reviewed as well as emerging alternatives, including diode lasers and the combined use of a special primary source, emitting a continuum, with a high-resolution spectrometer for atomic absorption spectrometry (AAS). Next, a general view of atomizers and their characteristics are also given. After revising the most popular wavelength selectors and their requirements for AAS, the commonly employed detectors used in ultraviolet–visible spectrochemical instruments are discussed.

As spectral background is a general problem to be corrected in order to obtain a proper AAS signal, a discussion on how the background is distinguished and subtracted from the total absorbance measured in the spectrometer follows. The chapter ends with a description of the main background correctors: deuterium background corrector, Zeeman correction, and Smith–Hieftje correction, implemented in commercial atomic absorption instruments.

3.1 INTRODUCTION: SINGLE-BEAM AND DOUBLE-BEAM INSTRUMENTS

An atomic absorption spectrometer consists basically of three main components: a light source, a sample cell in which gaseous atoms are produced, and a means of measuring the specific light absorbed. Figure 3.1a shows the simplest

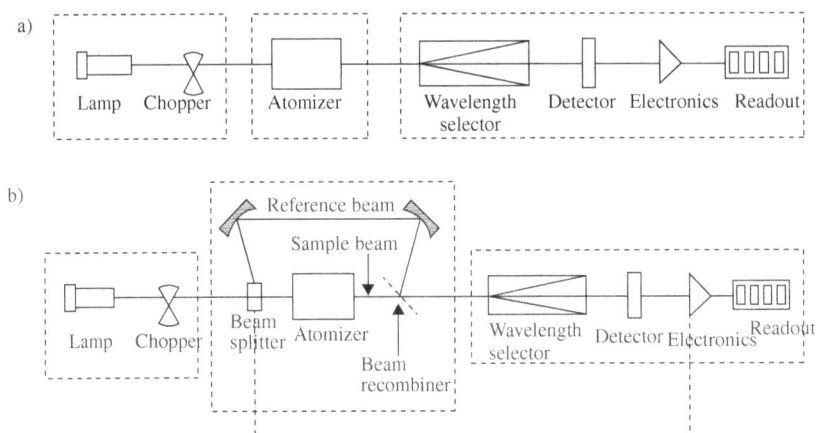

Figure 3.1. Instruments for atomic absorption spectrometry. (a) Single-beam spectrometer and (b) double-beam spectrometer.

configuration, called a "single-beam spectrometer," of such an instrument: the lamp (emitting intense and narrow lines), the atomizer (where the species of interest are converted into atoms), and the wavelength selector (e.g., a mono-chromator). To select a specific wavelength of light that is absorbed by the analyte, these are all aligned in a row. The light wavelength selected is focused onto a detector (e.g., a photomultiplier tube [PMT]), producing a signal pro-portional to the incident light intensity. The combination of the wavelength selector with a photoelectric detector for the isolated wavelength band(s) is usually called a spectrometer. However, it is important to note that it is com-mon in absorption spectrochemical measurements for the term spectrometer to be used when referring to the whole instrumentation setup including also the lamp and the sample cell.

To avoid detecting the atomizer continuum emission, the radiation source is modulated in order to provide a means of selectively detecting the ampli-fied modulated light coming from the source lamp while continuous emission from the sample cell is undetected. Source modulation can be accomplished with either a rotating chopper (mechanical modulation) located between the light source and the atomizer, or by pulsing the source with a pulsed power supply (electronic modulation). A synchronous detection eliminates the unmodulated direct current (DC) signal emitted by the atomizer from being measured and so only the amplified AC (modulated) signal coming from the lamp is measured.

The intensity of the light source may not remain constant during an analysis time. If only a single beam (Figure 3.1a) is used, a blank read-ing containing no analyte would have to be taken first, in order to set the absorbance detector value to zero. If the intensity of the source changes by the time the sample is put in place, the measured absorbance will be inaccurate. To obviate such inconvenience, an alternative instrument configuration is one based on a spectrometer that incorporates a beam splitter so that one part of the beam passes through the sample cell and the other is a reference. A diagram of a "double beam spectrometer" configuration is illustrated in Figure 3.1b. In double-beam instruments there is a continuous monitoring between the reference and sample beams. To ensure that the spectrum does not suffer from loss of sensitivity, the beam splitter is designed so that a proportion, as high as possible, of the total energy of the lamp beam is passed through the sample.

In the following sections of this chapter the choices for the basic components of an atomic absorption spectrometer will be considered: lamps, atomizers, wavelength selectors, and detectors. It is important to note that any radiation from the light source absorbed or scattered by other atoms or molecules, apart from the analyte, will give rise to a background absorption that will be summed up to the recorded specific absorption of the analyte. In the last section of this chapter, the means for measuring and correcting for this background absorption will be dealt with.

3.2 PRIMARY RADIATION SOURCES

The most common lamps used in AAS are sources emitting a narrow-line spectrum from the element of interest. This line should be of sufficient spectral purity and intensity to achieve a linear calibration graph and have a low level of baseline noise. If the intensity of the spectral line is too low, the noise performance of the instrument will be compromised by excessive photon shot noise (this arises from the random generation of electrons in the ultraviolet-visible [UV-Vis] light detector). On the other hand, if there is unspecific nonabsorbable radiation within the spectral band pass of the wavelength selector, the calibration curve will curve toward the concentration axis, leading to a loss of sensitivity at a high absorbance value. Such nonabsorbable radiation can result from the presence of another spectral line within the spectral band pass of the monochromator or from a continuum background radiation.

The main sources used for atomic absorption spectrometry (AAS) are the hollow cathode lamp (HCL) and the electrodeless discharge lamp (EDL). The HCL is a bright and stable line emission source commercially available for many elements. However, for some volatile elements (where low emission intensity and short lamp lifetimes are commonplace), EDLs are used; this is the case, for example, for As, Se, Hg AAS lamps. Boosted HCLs aimed at increasing the output from the HCL are also commercially available.

Emerging alternative sources, such as diode lasers and the combined use of a primary source emitting a continuum with a high-resolution spectrometer will also be described below.

3.2.1 Hollow Cathode Lamps

A schematic diagram of a typical HCL is shown in Figure 3.2a. It consists of a hollow cathode (containing the element of interest) and an anode. These are sealed in a glass tube filled with an inert gas at low pressure. By applying a potential difference of about 300–400 V between the anode and the cathode, the ionization of gas occurs and a discharge between the two electrodes takes place at the low pressure. Positively charged gas ions (e.g., Ar^+, Ne^+) are accelerated toward the cathode by the potential existing in the discharge. Under ion impact with the solid, metal atoms from the cathode are ejected, in a process called *sputtering* (see Figure 3.2b), to the plasma existing at the mouth of the cathode. Once in the plasma, sputtered metal atoms may collide with other high-energy particles, resulting in a transfer of energy and causing the metal atoms to become excited. Since this excited state is not stable, the metal atoms relax to their ground state, emitting radiation at the characteristic wavelengths of the element. In this way, more than one analytically useful spectral line can be generated for most elements.

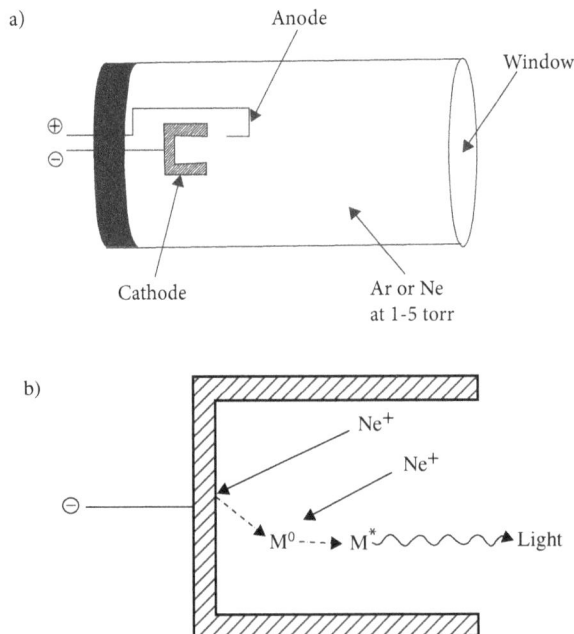

Figure 3.2. The hollow cathode lamp. (a) Diagram of a hollow cathode lamp and (b) sputtering, excitation, and emission processes.

The cylindrical shape of the cathode makes it easier to concentrate the radiation into a beam that passes through a transparent window.

3.2.1.1 Details of the Components of a HCL

Anode and cathode: An HCL contains a tungsten or zirconium anode and a cylindrical hollow cathode made from (or containing) the element to be determined. If the metal is stable in air and has a high melting point, the pure metal may be used (e.g., cathodes for lamps of Cu, Fe, Ni, and Al are usually machined from the corresponding solid metal, while hollow cathodes for expensive metals such as Pd, Au, and Ir usually have sheet metal inserts). If the metal is too brittle, a sinter of pressed metal powder is employed (e.g., Mn, W). If the metal is reactive in air, then the metal oxide or halide can be used (e.g., Na). A metal with a relatively low boiling point or a relatively high vapor pressure (e.g., Hg, Cd, Pb, Zn) is usually alloyed with another metal. The powder technique is employed for producing multi-element lamps where two or more metals are present (see Section 3.2.1.3).

Gas fill: The filler gas must be monoatomic to avoid molecular continuum spectra. Either high-purity neon or argon (1–5 torr) is used as the discharge gas. Generally, argon is only used when a neon spectral line would interfere with the wavelength for the selected element. The preferential use of neon is due to its higher ionization potential, which produces greater signal intensity. Helium is not used due to its low mass number making it an inefficient sputterer; besides, helium gives a short lamp lifetime due to a rapid lowering of the filler gas pressure by adsorption of gas atoms onto surfaces within the lamp.

Envelope: The electrodes are enclosed within a glass envelope with a quartz, or a special borosilicate, with an end window attached to it. For elements that emit at wavelengths of less than about 300 nm, quartz must be used, while for higher wavelengths borosilicate is typically used.

3.2.1.2 HCL Operation

HCL electrical current is the key parameter affecting analytical AAS results. An increase in the lamp current produces an increase in the intensity of the lamp element-specific light emission. However, as the operating current increases above the recommended value, increased broadening of the emission lines occurs and the phenomenon of self-absorption is observed at comparatively high HCL currents. In the presence of temperature gradients, an inversion of the peak top of the emission line (due to a cooler cloud of atoms in front of the cathode absorbing the cathodic emission within the lamp itself) may occur. Such a self-reversal effect gives rise to reduced sensitivity and, in some cases, a shorter linear interval in the calibration curve. However, it is worth mentioning that the temporary use of lamp currents higher than recommended may be advantageous for measurements near the detection limit.

From a practical point of view it should be ensured that there is a good stability of the HCL signal. Typical HCLs require a warm-up period (some minutes) after switching on, to stabilize their light output. Warm-up time is particularly important for single-beam instruments (the change in intensity of the lamp is reflected in the baseline of the instrument). With double-beam instruments, the need for lamp warm-up time is not so evident since the instrument is able to compensate for changes in the sample beam intensity since there is a continual comparison with the reference beam. Nevertheless, it is desirable to allow a short warm-up before attempting precise analytical measurements; this is because the profile of the emission line from the lamp can change during this period and so small changes in analytical signal may result.

Even the most carefully handled lamp will eventually fail to operate. As the filling gas is absorbed into the internal surfaces of the lamp glass, in time, the pressure of the fill gas will fall to a level that can no longer sustain the hollow cathode discharge. Operating the HCL at excessive lamp currents will accelerate this process. Furthermore, attempts to run a lamp at extreme currents can cause the cathode to overheat, and this can damage irreversibly the cathode (this is especially serious for the more volatile elements).

For some marketed HCLs it can be observed that brand-new lamps look like they have already been used (see Figure 3.3). Two treatments during processing in the lamp production step account for this:
- Some manufacturers heat treat the cathode material under vacuum to ensure that all absorbed gases are removed. During this purification stage (aimed to ensure a dependable performance), a layer of the cathode material is deposited on the inside of the glass envelope of the lamp. The amount of material deposited varies, depending on the volatility of the cathode element.
- Furthermore, a characteristic black patch on the lamp envelope near the anode can sometimes be seen, which is produced by subjecting the zirconium anode to ion bombardment. This step vaporizes a small amount of anode material and deposits it onto the lamp envelope producing the black spot. This zirconium metal film is highly reactive and acts as a very efficient getter of traces of oxygen and other impurity gases that might otherwise reduce the lifetime of the lamp. Therefore, the black getter spot close to the anode helps to prolong the useful life of the lamp and ensures continued spectral purity throughout the life of the lamp.

3.2.1.3 Multi-element HCLs

The HCL described earlier is designed to emit the atomic spectrum of a desired single element. Thus, a specific lamp should be selected for each element to be

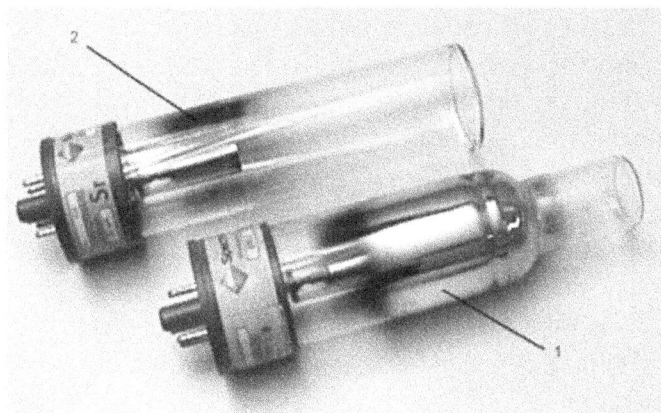

Figure 3.3. Photographs of some brand-new lamps. The thin layer of cathode material deposited onto the inside surface of the glass envelope (1) is noticeably different from the black getter spot (2).

determined. However, multi-element HCLs are also commercially available. Durable and commercially interesting multi-element HCLs need to fulfill requirements such as the following:

1. Spectral interferences should be absent. Spectral interference is the overlap of an emission line of the analyte with the emission line of another element(s) (interferent). For example, once-proposed Ag/Au lamps failed in this regard, since emission lines from silver interfere with the analytical line of gold and, in turn, gold emission lines interfere with the silver analytical line.

2. The chosen elements should be compatible, that is, they should coexist in the same cathode without undesirable effects restricting the lifetime of the lamp.

3. The selected analytical line for each element has to provide enough intensity while maintaining a low baseline noise.

4. Last but not least: the multi-element HCL must satisfy a market need. For example, a Ag/Cd/Pb/Zn lamp is useful for the environmental analysis market, while the Co/Cr/Cu/Fe/Mn/Ni lamp meets the needs of the base-metal market.

3.2.2 Electrodeless Discharge Lamps

For many elements an HCL is a most satisfactory spectrochemical source for atomic absorption. For volatile elements, such as As and Se, the low intensity and short lamp lifetime are a problem when using HCLs. For such elements, EDLs offer recognized advantages, in terms of useful lamp lifetime and higher light intensity.

The basic design of an EDL is shown in Figure 3.4. A small amount of the metal or salt of the element for which the source is to be used is sealed inside a quartz bulb that contains argon at low pressure. A radiofrequency (or microwave) coil surrounds the bulb. When power is applied, an intense radio-frequency (or microwave) field is created. The argon within the tube ionizes and gains kinetic energy from the field forming a low-pressure plasma. The plasma energy is transferred to the vaporized metal upon collisions. Finally, the excited metal vapor returns to its ground state by emitting light. An accessory power supply is necessary to operate EDLs with most spectrometer models.

EDLs are about ten times more intense than HCLs, but they usually suffer from unstable output. EDLs offer the analytical advantages of better precision and lower detection limits if the determination is intensity limited. EDLs are available for a wide variety of elements, including arsenic, antimony, selenium, mercury, cadmium, bismuth, and phosphorus.

3.2.3 Boosted Discharge Lamps

Boosted discharge lamps are HCLs in which a second discharge, electrically isolated from the sputtering discharge, is used to further excite the sputtered atoms. In a conventional HCL the only way to increase the intensity of the emitted light is to increase the lamp current. This causes more atoms to be sputtered leading to curvature of atomic absorption calibration graphs because the radiation from excited atoms in the cathode can interact with unexcited atoms on its way out of the cathode, giving rise to self-absorption and self-reversal of resonance lines.

In boosted discharge lamps the excitation process is made independent of the sputtering process, thus allowing a more efficient excitation of the sputtered atoms. In this way, the radiation does not have to pass through a cloud of ground-state atoms on its way out of the lamp; consequently, there is less self-absorption and the calibration graphs are less curved. A preferential excitation of the lines of analytical interest is observed in many cases, although,

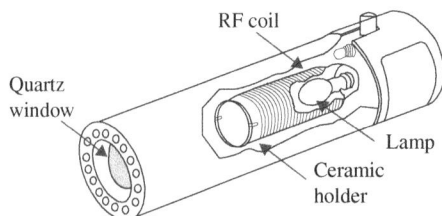

Figure 3.4. Schematics of an electrodeless discharge lamp. Reproduced with permission from Perkin Elmer.

boosted discharge lamps are not equally effective for the interesting analytical lines of all elements. In some cases, little or no increases (and even for some particular elements a reduction) in intensity of the corresponding analytical line have been observed in a boosted HCL.

3.2.4 Diode Lasers

The potential of diode lasers as sources of resonance radiation for AAS has been intensively investigated in recent years. Single-mode diode lasers are proposed for AAS as tunable narrow band-line sources, offering a number of attractive features:

1. The power of commercial diode lasers is between 1 and several orders of magnitude higher than that provided by HCLs. In addition, diode lasers show good signal stability.
2. In contrast to the above-described line sources, a diode laser emits a prominent single line, under normal operating conditions, which simplifies the spectral isolation of the absorption signal.
3. The typical line width of a commercial diode laser is approximately two orders of magnitude lower than the width of absorption lines in flames and furnaces. This allows the expansion of the linear dynamic range of the calibration curve to high concentrations of an analyte, by detection of absorption in the wings of the absorption line.
4. The spatial coherence of diode lasers makes it possible to deliver a narrow laser beam from a distance without noticeable divergence and to easily manipulate the spatial profile.
5. The wavelength of diode lasers can be easily modulated at frequencies up to GHz by modulation of the diode current. Wavelength modulation of the diode laser with detection of absorption at the second harmonic of the modulation frequency greatly reduces low-frequency noise (flicker noise) in the baseline and provides improved detection limits.

Unfortunately, perhaps the main drawback of current diode lasers for AAS is the narrow spectral range of the commercially available diode lasers, which limits its applicability when compared with HCLs.

3.2.5 Continuous Sources

Line sources, such as HCLs and to a lesser extent EDLs, have been used almost exclusively for routine application of AAS. Their stable, narrow-line emission at the center of the absorption profiles (as proposed by A. Walsh originally) guarantees high analyte specificity and good detection limits, even with low-resolution monochromators. Conversely, the use of a source emitting a continuum will require a resolution as high as approximately 2 picometers.

The present availability of continuum sources with a high emission intensity within the spectral interval of interest, along with high-resolution echelle spectrometers and solid-state arrays detectors, have prompted the exploitation of the advantageous use of continuous-spectrum light sources in AAS. A high-intensity continuous source (a xenon short-arc lamp) in combination with a high-resolution double-echelle monochromator and a linear charge-coupled device (CCD) array detector, has been commercialized for AAS measurements with a bandwidth of about 1.6 pm per pixel at 200 nm. The instrument allows for multi-elemental analysis, determination of elements for which reliable line sources are not available, and also simultaneous background correction.

Figure 3.5 shows the experimental setup for such an AAS arrangement using an electrothermal atomizer.

3.3 ATOMIZERS: A GENERAL VIEW

Two systems are most frequently used in AAS to produce atoms from a *liquid* or *dissolved* sample:

1. A flame, where the solution of the sample is aspirated. This system is addressed in Chapter 4 of this book, which is dedicated to flame atomic absorption spectrometry.

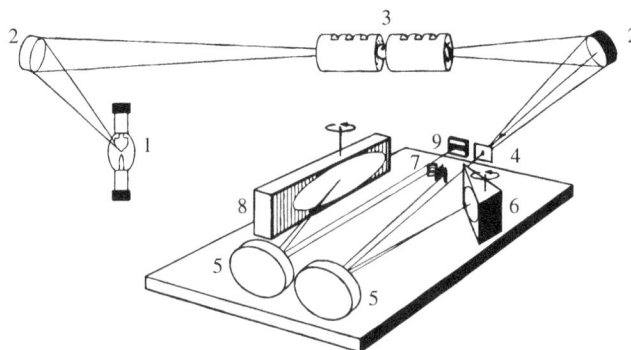

Figure 3.5. Experimental setup for continuous-source atomic absorption spectrometry with double echelle monochromator. (1) Xenon short-arc lamp, (2) off-axis ellipsoid mirrors, (3) longitudinal Zeeman graphite furnace module, (4) entrance slit, (5) off-axis parabolic mirrors, (6) Littrow prism, (7) deflection mirrors and intermediate slit, (8) echelle grating (76 grooves/mm blaze 76°), and (9) linear CCD array detector. Reprinted with permission from Heitmann et al. (1996) © by Elsevier.

2. An electrothermal atomizer, where a drop of the liquid sample is placed into an electrically heated graphite tube. Chapter 5 focuses on electrothermal atomization—atomic absorption spectrometry.

It is interesting to note that some commercial instruments now have pre-aligned flame and furnace atomizers, making it feasible to select either of them purely by software means.

Gaseous and volatilized analytes can also be easily determined by AAS using either flame or electrothermal atomization. The determination of several elements by formation of covalent volatile hydrides (e.g., arsenic, selenium) and cold vapor generation (mercury and cadmium) will be reviewed in Chapter 6. In addition, in the last sections of Chapter 7, the coupling of gas chromatography to such atomizers is explained.

Other atomizers have been investigated for atomic absorption spectrometric measurements apart from flame and electrothermally heated cells. For example, a jet-assisted glow discharge system has been commercialized for the direct analysis of *solid* materials. The source can be used for both bulk analysis of homogeneous solids and for depth profiling of layered materials. The atomization cell is a vacuum chamber with two electrodes (the sample acts as the cathode). When the sample is in position, the chamber is rapidly evacuated and argon gas flows through the chamber. An electrical current ionizes the argon, which bombards the cathode surface causing sample sputtering. The resultant sputtering is an efficient means of atomization. The free atoms are entrained in a cohesive atom cloud into a space along the axis of the discharge chamber where absorbed light (from an external lamp) can be measured, as usual.

3.4 WAVELENGTH SELECTORS

Wavelength selection in UV-Vis spectrochemical instruments can be based on filters, spatial dispersion of wavelengths, or interferometry. Wavelength selectors that spatially disperse the spectral components of an optical beam are the most common. Basically, they consist of an entrance slit (which defines the area of the source of radiation that is viewed), at least one dispersive element that can be a prism or a grating, an image transfer system (mirrors and lenses), and a focal plane with one or several exit slits.

Most wavelength selectors used in AAS are monochromators. In such dispersive wavelength selectors, an exit slit about the same size as the entrance slit is used to isolate a small band of the spatially dispersed wavelengths that are incident on the focal plane. One wavelength band at a time is isolated and different wavelength bands can be selected sequentially by rotating the dispersive element in order to bring a new band into the proper orientation so that it will pass through the exit slit. The *linear dispersion*, D_l, selects how far apart in distance two close wavelengths are separated in the focal plane. D_l is conveniently expressed in mm.nm^{-1} units. Most often it is the *reciprocal linear dispersion*, R_D, that is specified for a monochromator. R_D represents the number of wavelength intervals (e.g., nm) contained in each interval of distance (e.g., mm) along the focal plane.

The *spectral band width* (SBW) or spectral band pass is the half-width of the wavelength distribution passed by the exit slit. Except at small slit widths, where aberrations and diffraction effects must be considered, the SBW is controlled by the monochromator dispersive element and the slit width. The attained SBW affects the spectral isolation of the analytical line and the required SBW is normally determined by the nearest line adjacent to the analytical one in the spectrum.

Figure 3.6 shows the configuration of four types of monochromators used in commercial AAS instruments. Figure 3.6a shows schematically a grating monochromator based on the Czerny–Turner configuration. Incident radiation passes through the entrance slit and strikes a parabolic mirror (collimating mirror). The entrance slit, at the focal point of the collimating mirror, acts as a point source for the collimating mirror, which produces parallel radiation for the dispersion element (a grating). The grating spatially disperses the spectral components of the incident radiation. Collimated rays of diffracted radiation strike a parabolic focusing mirror that focuses dispersed radiation in the focal plane. The exit slit, placed in this focal plane, isolates a particular narrow wavelength interval.

The Ebert type (Figure 3.6b) is very similar to the Czerny–Turner type except that one large mirror serves as both the collimator and the focusing mirror. The Czerny–Turner system has the advantage that focussing two mirrors is usually easier than focusing the single mirror of the Ebert monochromator. Also, two smaller mirrors are less expensive than one large mirror. Figure 3.6c shows the Littrow configuration. As can be seen, the Littrow configuration is that specific geometry in which the light of a specific wavelength diffracted from a grating, into a given diffraction order, travels back along the direction of the incident light (the deviation angle is close to zero). In some cases, a beam splitter is used to direct the diffracted radiation toward a more convenient focal plane in order to avoid having the entrance and exit slits too close.

Finally, Figure 3.6d shows an echelle configuration. Originally invented for astronomy, echelle optics are now widely used in atomic spectrometry. It is a very compact optical design, incorporating usually a coarse grating (e.g., 120 grooves/mm) and a prism as a cross-disperser, in a Czerny–Turner configuration. There are many designs of echelle optics, most using prisms but some using a grating as the cross-disperser, although the basic format is similar.

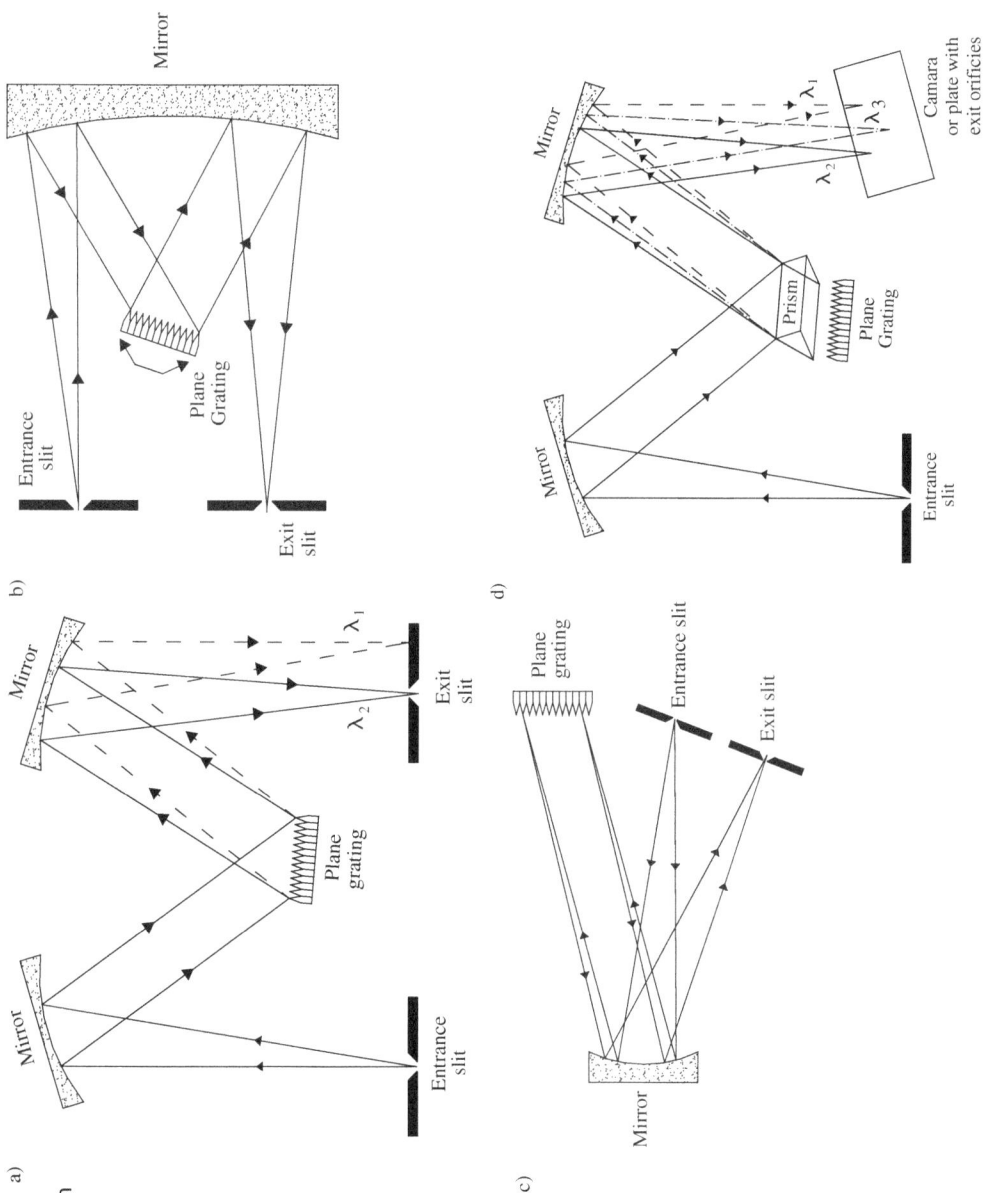

Figure 3.6. Dispersive wavelength selectors. (a) Czerny turner; (b) Ebert; (c) Littrow; and (d) echelle.

The prism in the echelle monochromator serves as an order sorter. The coarse grating generates low resolution and different diffraction orders are dispersed in the same direction. Then the prism disperses the spectrum perpendicular to the diffraction direction, giving rise to enhanced resolution in a two-dimensional spectrum (the diffraction orders are separated by the prism). Instead of having a spectrum with intensity of every line versus wavelength (i.e., one dimension), the echelle configuration creates a two-dimensional spectrum with wavelengths along a horizontal line and a series of lines (order of the spectra) going down (like the text in a page of a book). In recent years, echelle optics are being incorporated into commercial AAS instruments for multi-element analysis in combination with an auto-aligning turret (e.g., with eight HCLs). Furthermore, the advantages of using an echelle configuration in combination with a continuous primary source are obvious, since the details of the spectral background close to the line can be measured with a wider exit slit and a linear multichannel detector. The use of such a two-dimensional multichannel detector allows for true simultaneous multi-elemental analysis.

3.5 DETECTORS

The intensity of the light passing through the exit slit of a wavelength selector must be quantified with a proper transducer (see Figure 3.1). In this section the most commonly used photon detectors will be briefly discussed.

The classical *vacuum phototube* consists of a photosensitive cathode and an anode in an evacuated glass or quartz enclosure. A high voltage is set between the two electrodes. Photoirradiation of the cathode causes photoelectrons to be emitted and attracted to the anode, causing current to flow: a flow which can be amplified and measured. Phototubes are used to detect moderate light levels.

In atomic spectrometry, however, the related and more sensitive PMTs are preferred to vacuum phototubes. A *PMT* can be considered as a vacuum phototube with additional amplification by electron multiplication. It consists of a photocathode, a series of dynodes as a chain on which secondary-electron multiplication occurs, and an anode. Photons strike the photocathode, which emits electrons due to the photoelectric effect. Instead of collecting these first few electrons (there should not be a many, since the primary use for PMTs is in low-intensity applications) at an anode, like in a phototube, the electrons are accelerated toward a series of additional electrodes, called *dynodes*, each at a more positive potential (50–90 V) than the preceding one. Additional electrons are generated at each dynode. This cascading effect creates 10^5 to 10^7 electrons for each photon hitting the photocathode. This amplified signal is finally collected at the anode where its intensity can be measured.

Figure 3.7 shows a schematic diagram of a PMT. According to the desired characteristics in terms of response time, gain, and so forth, different types of photocathode and dynode structures have been developed. The PMT generally has the photocathode in either a side-on or a head-on configuration. The side-on type receives incident light through the side of the glass bulb, while the head-on (or end-on) type receives light through the end of the glass bulb. Figure 3.8 shows the external appearance of both configurations. The head-on type has a semitransparent photocathode (transmission mode photocathode) deposited upon the inner surface of the entrance window. Most side-on types

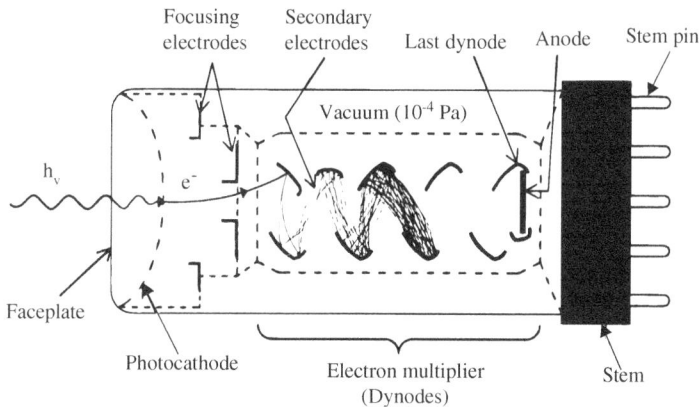

Figure 3.7. Schematic diagram of a photomultiplier tube.

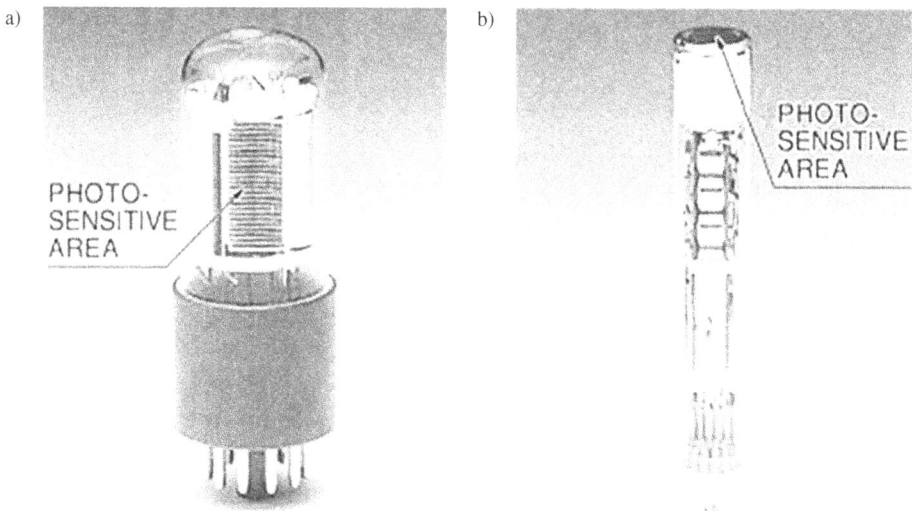

Figure 3.8. External appearance of photomultiplier tubes. Reproduced with permission from Hamamatsu. (a) Side-on type and (b) head-on type.

employ an opaque (reflection mode photocathode) and a circular-cage structure electron multiplier.

The interesting properties of the PMT are numerous: large wavelength coverage, large dynamic range, high amplification gain, and low noise. However, a PMT is a single detector, and for this reason, there is a current trend toward the replacement of PMTs by multichannel detectors. These consist of an assembly of many individual, small adjacent detectors called pixels. Charge-transfer device detectors are currently used, and are based on CCD or charge-injection device technology. An assembly of linear arrays or a two-dimensional assembly can be used as the detector at the exit slit of the wavelength selector (in such a case, the aperture of the exit slit is larger than that when a PMT is used).

3.6 BACKGROUND CORRECTORS

Contributions to the background of a measurement in AAS can arise from spectral interferences due to a spectral line of another element within the band pass of the wavelength selector (such a possibility is rather uncommon in AAS; besides, such spectral interferences are now well characterized), absorption by molecular species originating from the sample, and light scattering from solid or liquid particles present in the atomizer.

To obtain the accurate absorbance due to the analyte it is necessary to subtract the background from the total absorbance measured in the spectrometer. Some commercially available strategies will be discussed below. It is important to keep in mind that the ideal for background correction should be a measure of a true blank solution.

3.6.1 Deuterium Background Corrector

A common, rather inexpensive, background correction approach is based on the use of a continuum source. A deuterium lamp is the radiation source used to correct for background absorption and consists of a deuterium-filled discharge lamp that emits a continuum spectrum from 180 nm to about 400 nm, with a maximum near 250 nm. This is the spectral region in which most analytical atomic absorption lines occur and where the effects of background absorption are most pronounced.

The deuterium lamp consists of a heated, electron-emitting cathode, a metallic anode, and a restrictive aperture between the two. A discharge current of several hundred milliamperes excites the deuterium gas, emitting a continuum. The discharge is forced to pass through a small aperture, forming a defined area of rather high light emission. A suitable window allows light transmission to the spectrometer's optical system.

Figure 3.9a shows the optical configuration of a single-beam spectrometer with deuterium background correction. In this configuration, the HCL and the deuterium lamp are sequentially pulsed on and off. When the HCL is on and the D_2 lamp is off, the total absorbance at the analytical wavelength is measured. This total absorbance comprises the absorbance from the analyte and the absorbance from the background at that particular analytical line. When the HCL is off and the D_2 is on, the D_2 continuum energy fills the slit (see Figure 3.9b), and so the average of the background absorbance across the whole slit is measured. Since the spectral band pass of the wavelength selector is usually large (e.g., 0.2 nm) compared to the analyte atomic line half-intensity width (less than 0.005 nm), the absorbance of the analyte is negligible compared to that of the average background when the deuterium lamp is used. Thus, the AAS true signal is calculated by subtracting the average

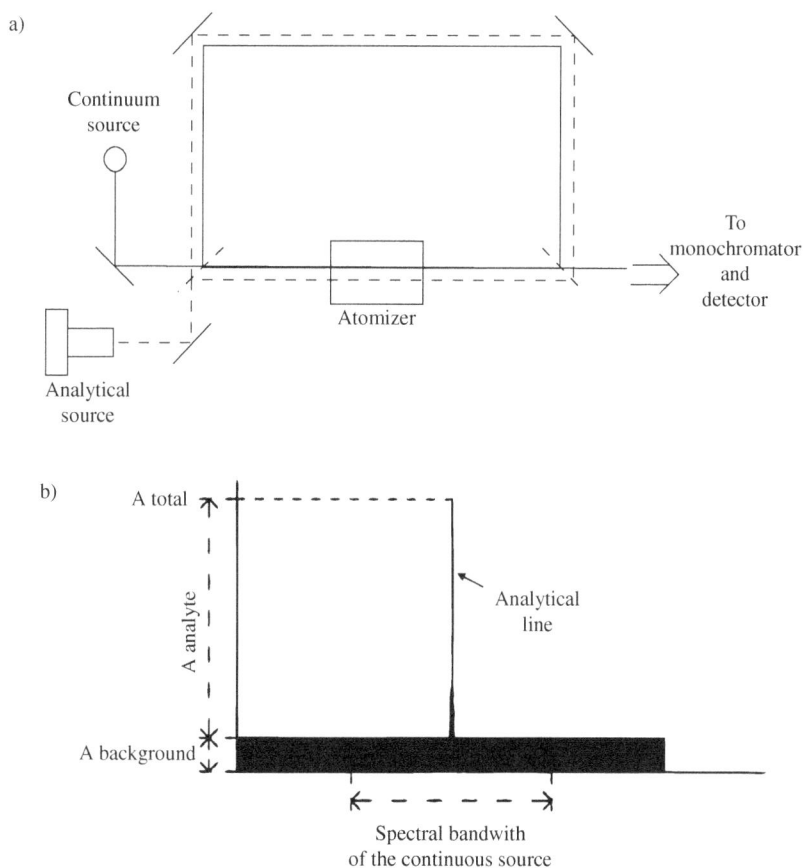

Figure 3.9. Deuterium background corrector. (a) Schematics of a single-beam AAS spectrometer with deuterium background correction. (b) Basis of the correction. A_{total}: total absorbance at the analytical wavelength. $A_{analyte}$: absorbance due to the analyte. $A_{background}$: absorbance at the analytical wavelength produced by the background.

of the background absorbance (deuterium lamp) from the total absorbance (measured with the HCL).

It is important that both the deuterium source and the HCL are well aligned so that their radiations follow the same optical path. If they are not, then the two measurements may not be made on the same atom population and significant errors may occur. To obtain successful background correction, the intensity of the deuterium lamp must be matched to that of the HCL. For example, an approach to follow if the light from the continuum source is too intense consists of reducing the spectral bandwidth (the energy measured from the continuum source increases with the square of the spectral bandwidth, while the radiant energy measured of the atomic spectral line from the HCL increases linearly with the spectral bandwidth); conversely, the SBW should be increased if the HCL intensity is too high compared to that of the deuterium lamp.

The deuterium background corrector can neither accurately correct for high absorbances nor for a structured molecular background (since the average of the background absorbance measurements may not be representative of the actual background at the analytical line).

3.6.2 Zeeman Correction

The splitting by a magnetic field of the spectral lines of an atom, as well as the polarization of those lines, constitutes the basis of background correction by the Zeeman effect.

Principles of Zeeman Effect

The quantum states of an atom undergo drastic changes when that atom is placed in a magnetic field; energy states that were "degenerate" may separate from each other and so spectral lines may split into three or more components. Transitions between these new states are given by the usual selection rule:

$$\Delta M_j = 0, \pm 1$$

where Mj is the magnetic quantum number.

The $\Delta M_j = 0$ components (called the π components) have their electric vectors linearly polarized parallel to the magnetic field, while the $\Delta M_j = +1$ components (called the $\sigma\pm$ components) have their electric vectors linearly polarized perpendicular to the magnetic field. The $\sigma\pm$ components are circularly polarized about the magnetic lines of force, also, with the σ^+ y σ^- component vectors rotating in opposite directions.

The "normal" Zeeman effect (called normal because it was explicable by classical physics) arises when both states π and σ are split by an equal amount of energy. A triplet is then obtained with a central π component and two σ^- and σ^+ components (see Figure 3.10). In this case, the component separation is directly proportional to the

---

magnetic field. The "anomalous" Zeeman effect is more complex and leads to the formation of a multiplet whose separation is a more complicated function of the magnetic field.

Figure 3.11 shows a diagram of an optical system of a commercialized AAS with an alternating current (AC) transverse (magnetic field applied across the light path) Zeeman corrector. As can be seen, a modulated electromagnet is located around the atomizer (a graphite furnace). A polarizer is positioned between the atomization cell and the wavelength selector, in order to remove the π component of the transmitted radiation. When the field is off, the total absorbance is measured (analyte and background), while when the field is on only the background is measured. Therefore, the background measurement can be made at the exact wavelength when the magnetic field is applied.

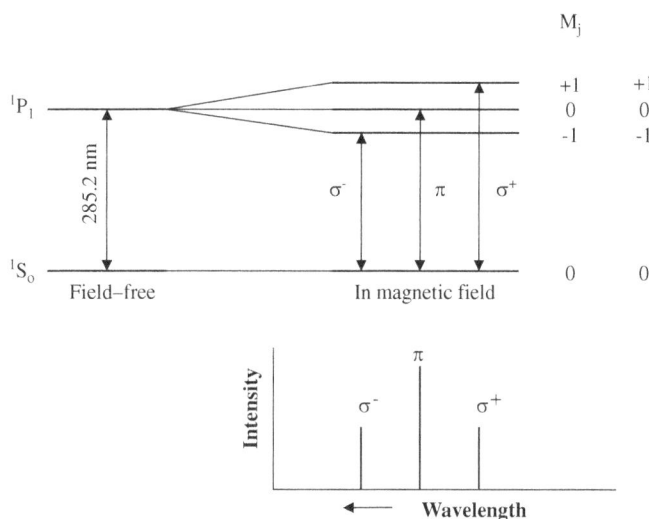

Figure 3.10. Normal Zeeman effect in the 285.2 nm magnesium line.

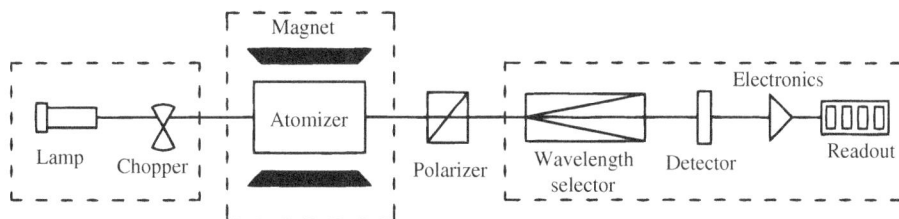

Figure 3.11. Diagram of the configuration of an AAS with Zeeman background correction.

In some configurations the polarizer is located prior to the atomization cell and a permanent magnet is used. By alternately polarizing the incident light parallel and perpendicular to the magnetic field, the background absorption is measured first and then the background plus sample absorption is measured. This is so because the *s+* components have been shifted away from the sample atom's resonance line and the remaining π components of the resonance line absorb only parallel polarized light, while the background equally absorbs both polarizations of light.

All Zeeman systems are optically single-beam spectrometers and require only a single energy source. Although the magnetic field could be applied to the lamp (direct Zeeman effect), most commercially available AAS systems apply the magnetic field to the atomizer (inverse Zeeman effect). The magnetic field is characterized by its mode (transverse or longitudinal) and its frequency. DC Zeeman background correctors use a permanent magnet and a rotating or vibrating polarizer to separate the combined and "background-only" signals. AC systems use an electromagnet, and measure the combined and "background-only" signals by alternately turning the magnetic field on and off. An important difference between longitudinal (magnetic field applied along the light path) and transverse AC Zeeman systems is that transverse systems use a polarizer, while the longitudinal configuration does not require the polarizer.

One of the advantages of Zeeman background correctors is that they enable background signals to be measured at exactly the analytical wavelength. Therefore, structured molecular background and spectral interferences are easily corrected for. Besides, this strategy corrects for high levels of background absorption. A disadvantage of the Zeeman background corrector, however, is "calibration rollover." Calibration rollover occurs when the σ components of the analyte are not sufficiently shifted from the analyte line as a result of which an overlap occurs at the absorption line. Rollover can cause two different analyte concentrations to register as having the same absorbance. To prevent such rollover problems occurring in practice, maximum permissible absorbance should be known. Another possible disadvantage of the Zeeman effect is loss of sensitivity for some elements.

3.6.3 Smith–Hieftje Correction

In the Smith–Hieftje background correction method the HCL is pulsed alternatively at low and then at very high current (see Figure 3.12). The use of high currents during short periods of time does not give rise to melting problems (and therefore shorter lamp lifetimes) of the HCL since the average current is low.

During the low-current part of the cycle (see Figure 3.12) the line is narrow and the analyte and background are measured together. However, during

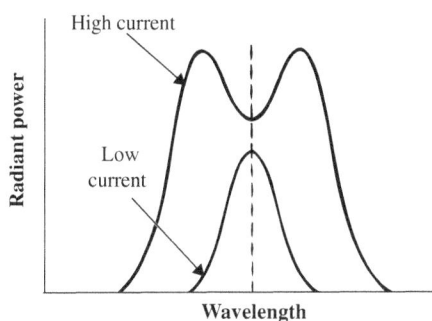

Figure 3.12. Emission line profiles in hollow cathode lamps operated at high and low currents.

the brief high-current pulse, the lamp emission lines broaden as a large cloud of atoms is formed in front of the lamp cathode. This cloud essentially prevents resonant radiation from reaching the analyte in the atomizer due to HCL emission self-reversal (see Chapter 1). In this part of the cycle, mainly the background is probed. The method is simple, it does not require additional components, and it can be used with any type of atomizer that is commercially available today.

4

Flame Atomic Absorption Spectrometry

Throughout this chapter, basic flame atomic absorption spectrometry (FAAS) is thoroughly described. The instrument components are first detailed individually and then are presented in an integrated manner. Common accessories such as autosamplers, microsamplers, high-solid analyzers, and so forth, are also discussed. Analytical figures of merit achieved with flame atomic absorption as well as potential risks of interferences (and how to overcome them) are dealt with in this chapter as well.

It ends with a description of four selected case studies (determination of calcium in milk; molybdenum in fertilizers; lead in gasoline; and boron, phosphorus, and sulfur for plant analysis); which show the broad field of applications of this widespread analytical technique.

4.1 INTRODUCTION

As described in previous chapters, for atomic absorption spectrometry (AAS) measurements, the analyte must be converted into free gaseous atoms, usually by the application of heat in the atomizer (e.g., using a flame or an electrically heated furnace). Analytical atomic spectroscopy based on the use of a flame as atomizer was first developed in the middle of the 20th century, then called flame atomic emission spectrometry (FAES), the precursor of flame atomic absorption spectrometry (FAAS). In 1964, with FAES and FAAS already enjoying a strong development, J.D. Winefordner in the United States and T.S. West in the United Kingdom demonstrated that atomic fluorescence in flames can be also used for analytical purposes.

Other atomizers offering higher analytical sensitivity than flame atomizers have been developed with great success, in particular, graphite furnace/electrothermal atomization in AAS. However, flame-based atomic instruments are still widely employed, especially for routine applications in inorganic elemental analysis. In Chapter 2, it was emphasized that flame-based atomic spectrometric instruments are extensively used mainly because they are very robust, are of comparatively low cost, corresponding analytical methods with FAAS are well established and validated, and the analyses are rapid. Presently, flame-based atomic techniques can be considered as a cost-effective choice whenever the expected concentrations of the analyte are in the mg/L level.

It should be stressed again here that flame methods based on atomic emission, atomic absorption, and atomic fluorescence are usually considered as complementary (see Chapter 2), but FAAS is by far the most commonly used, providing reliable results even when used by people with limited training.

4.2 THE ATOMIZER UNIT IN FLAME ATOMIC ABSORPTION SPECTROMETRY

Although methods have been described in which solid samples are introduced directly or as suspensions into a flame for AAS, the analysis of liquid samples and solutions is much more common (FAAS can be also used for volatile or volatilized analytes and Chapter 6 addresses this topic). As can be seen in Figure 4.1, a liquid sample containing the analyte, M, needs to undergo several conversion stages in order to obtain the desired maximum number of analyte atoms in the fundamental state once in the atomizer. In FAAS the liquid sample, in the first step, is converted into a fine spray or mist (this step is called nebulization). Then, the spray reaches the atomizer (flame) where a series of processes, including desolvation, volatilization, and dissociation, takes place to produce gaseous free atoms. As is illustrated in Figure 4.1, once the gaseous free atoms of M are formed, they can undergo other processes while in the flame, such as ionization, excitation, and chemical reactions. We will now explain each process in some detail.

Desolvation: Evaporation of the solvent is the first step that occurs after nebulization, leaving a dry aerosol of molten or solid particles. Water tends to lower the flame temperature; however, desolvation efficiencies can be quite high when larger droplets are removed prior to entering the flame (in the nebulization chamber). Organic solvents evaporate more rapidly than water and the desolvation can be accelerated by the heat of organic vapor combustion.

Volatilization and dissociation: The solid or molten particles present after desolvation must be vaporized to obtain gaseous molecules to be then dissociated to produce the sought-

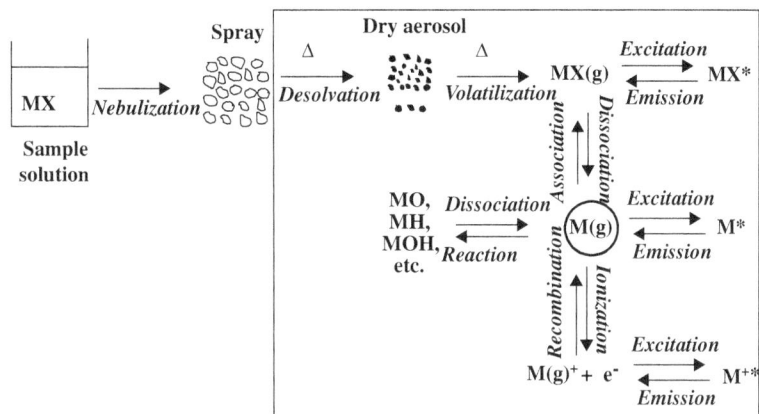

Figure 4.1. Diagram showing the steps involved in obtaining fundamental analyte (M) atoms from a liquid or dissolved sample containing the salt MX. (*) excited state.

after free atoms. Incomplete volatilization leads eventually to a loss of analyte free atoms and as a consequence analytical signal decrease. It could also cause nonlinearity in analytical calibration curves, continuous background emission from incandescent particles, and light scattering.

Ionization, excitation, and chemical reactions: In the vapor phase of a typical flame, the analyte can be present as free atoms, molecules, or ions. Molecular species, such as MO, MH, or MOH, can be formed by reactions of analyte atoms with gaseous flame constituents or with volatilized species from the sample (high temperatures and a reducing environment tend to reduce the formation of refractory oxides). The formation of molecular species and ions reduces the concentration of free atoms and thus adversely affects the detection limit (DL). As was explained in Chapter 2, ionic species are also formed by the loss of an electron from an atom. Therefore, for analytes with low ionization potentials, the use of a low-temperature flame and a spectroscopic buffer (readily producing electrons to retrograde the equilibrium) are advisable.

Finally, as is shown in Figure 4.1, in some cases, the flame can be efficient enough to excite gaseous atoms (e.g., alkalis) allowing for spontaneous emission of radiation characteristic of the analyte.

4.2.1 Nebulizer, Nebulization Chamber, and Burner

The most commonly used nebulization systems are pneumatic devices in which a jet of compressed gas (the nebulization gas) aspirates and nebulizes the solution. Figure 4.2 shows a picture and a diagram of a typical pneumatic concentric

a)

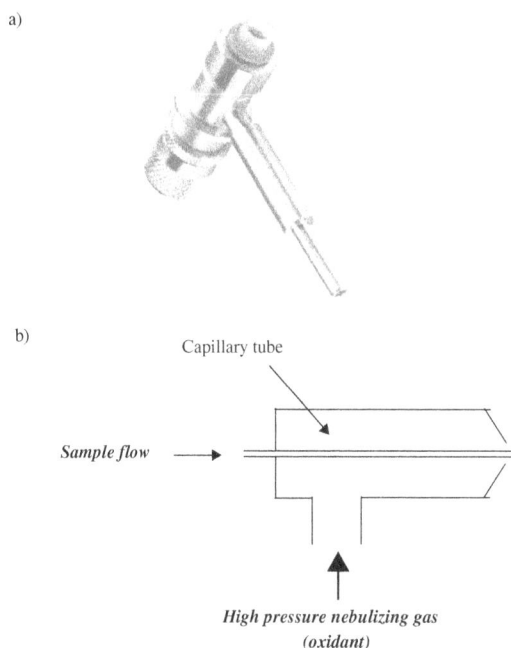

b)

Capillary tube

Sample flow →

*High pressure nebulizing gas
(oxidant)*

Figure 4.2. Nebulizer used in FAAS.
(a) Photograph of a typical nebulizer.
(b) Basic schematics of a venturi-based
nebulizer.

nebulizer used in FAAS. In such a device, the gas flows through a small open-ing that concentrically surrounds the capillary tube, causing a reduced pressure at the tip (Venturi effect) and a continuous suction of the sample solution from its container (giving rise to steady-state analytical signals). In most cases, the aspiration rate is proportional to the pressure drop along the capillary and inversely proportional to the viscosity of the solution. The sample solution, drawn up the capillary tube, encounters at the exit the high-velocity nebulizing gas, which causes the formation of droplets of various sizes. Pneumatic nebu-lizers produce droplet diameters typically in the range 1–50 mm.

Nebulizers are commonly made of stainless steel (preferred for general applications with acid concentrations less than 5%). However, for maximum chemical resistance, there are also available platinum alloy nebulizers (not suit-able for use with aqua regia or hydrofluoric acid) consisting of a platinum alloy (e.g., Pt–Ir) needle assembly and a venturi made of materials such as tantalum or poly(ether ether ketone) (PEEK). The connection between the nebulizer and the liquid sample is usually a plastic capillary tubing (about 15 cm long). Of course, it has to be always ensured that this plastic capillary tubing is cor-rectly fitted to the nebulizer capillary; any leakage of air, tight bends, or kinks will cause unsteady, nonreproducible readings.

At times the plastic capillary tubing can become clogged and it will be necessary to cut off the clogged section or fit a new piece of capillary tubing. The nebulizer capillary can also become clogged. If that occurs, the flame should be turned off and the nebulizer dismantled and placed into an ultrasonic cleaner containing liquid soap solution. If the ultrasonic bath treatment fails to clear the blockage, a burr-free wire should be carefully passed through the nebulizer and then the ultrasonic cleaning procedure repeated.

Two types of burners have been employed in flame spectroscopy: *total consumption burners* and *premix chamber burners*. In the *total consumption burner*, a concentric nebulizer without a spray chamber is used. A third concentric orifice transports the required fuel around the nebulizer tip so that the nebulizer is an integral part of the burner (see Figure 4.3). The flame is supported immediately above the nebulizer tip. The aspirated sample is all directed into the flame, without any droplet-size selection, allowing the introduction of relatively large amounts of sample into the flame. Some disadvantages include, noisy burners (both from the electronic, and auditory standpoints), a relatively short path length through the flame, and problems arising from tip clogging.

Presently, total-consumption burners in FAAS are rarely used. A commercial instrument commonly employs a *premix chamber* (see Figure 4.4). In the premix chamber, large droplets of the nebulized sample solution condense and drain out while the remaining fine droplets mix with the fuel and oxidant gases before they enter the burner head. Devices that remove large droplets usually

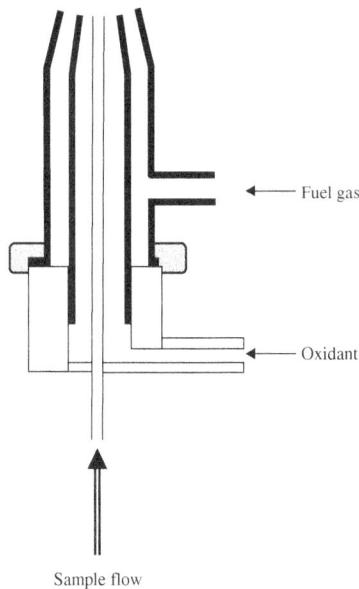

Figure 4.3. Schematic of a total-consumption burner.

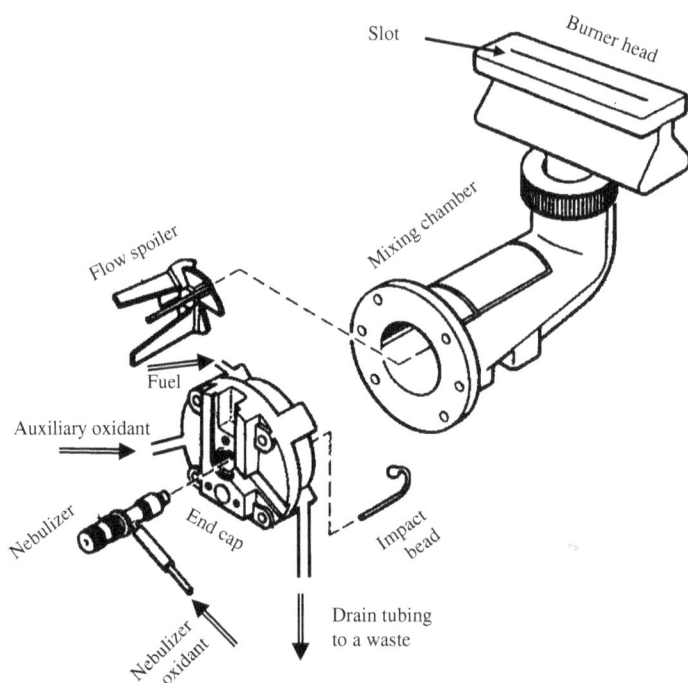

Figure 4.4. Example of a nebulizer, premix chamber (it can be used alternatively with either the flow spoiler or the impact bead), and burner head.

consist of a paddle (e.g., flow spoiler) or an impact bead. Typically, about 90% of the sample droplets condense, leaving only 10% of the sample to enter the flame. In flame atomizers, the nebulization gas is usually the oxidant (e.g., air), and the fuel is brought into the nebulization chamber through a separate port. The mixed oxidant, fuel, and finer sample mist are then carried to the burner head, which supports the flame. Although a large portion of the aspirated sample is lost in the drain of the premix chamber, the "atomization efficiency" of the portion of the sample entering the flame is greater than that in total consumption burners, because the droplets are much finer. Furthermore, the path length is longer. In addition, combustion with a premix burner provides less-noisy signals.

A drain vessel must be used to gather the effluent from the premix chamber drain. The drain vessel should be checked regularly and replaced when necessary. Therefore, the vessel should not be stored in an enclosed storage area. Rather, it should be stored in plain sight of the operator, usually on the front of the instrument or on an open shelf underneath the instrument table.

In Figure 4.4, a typical burner head of a premix burner is shown. As can be seen, the slot, which is usually aligned with the instrument's light path, is long and thin (typically several cm long and tenths of mm wide). Some manufacturers also market specially designed burners (e.g., a three-slot burner head) that do not clog easily even when concentrated solutions are used. Burner heads are made of all-titanium or of special corrosion-resistant alloys.

During aspiration of certain solutions, carbon and/or salt deposits can build up on the burner causing changes in the fuel/oxidant ratio and flame profile. To minimize the accumulation of salts, a dilute solution of acid may be aspirated between samples. However, if salts continue to build up, the flame should be turned off and an appropriate cleaning strip should be inserted in the burned slot and moved back and forth through the slot (care should be taken not to not use sharp objects such as razors to clean the burner as they can damage the slot and form areas where salt and carbon can accumulate at an accelerated rate). If such cleaning proves to be inadequate the burner should be removed and soaked in warm soapy water, or in dilute acid, or in an ultrasonic bath with non-ionic detergent.

4.2.2 Flame

Common flames used in AAS are today obtained with a premix burner where fuel and oxidant are mixed in a previous chamber before combustion in the burner. These "premix" flames are very stable and show a structure consisting of defined cones (see illustration in Figure 4.5). The appearance and the size of these regions depend on the type and ratio of the "fuel/oxidant" used. In the *inner zone* (primary combustion zone), the fuel/oxidant is ignited and combustion reactions proceed rapidly to completion or as far as the available oxidant allows. The *interconal zone* (which corresponds to the hottest part of the flame) lies immediately above. In that region, local thermal equilibrium exists and it is the region in which most analytical measurements are carried out. This interzonal region is externally surrounded by the *outer zone* (secondary

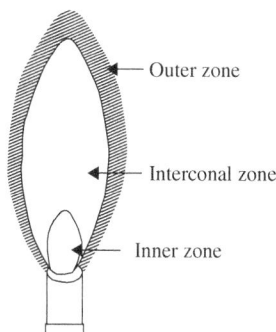

Figure 4.5. Typical flame regions.

combustion zone), which is much more diffuse and less well defined than the primary combustion zone. This external region is in contact with the atmosphere, so oxygen and nitrogen from the surrounding air can penetrate and cause additional reactions to take place within it.

The part of the flame to be measured in a given analysis depends on the analyte (generally the flame position is adjusted to yield a maximum absorbance reading). It is important to note that though the temperature of the flame is of great significance, the oxidizing or reducing characteristics of the flame are also of paramount importance. For example, Figure 4.6 shows the relative observed absorbance of three elements as a function of distance (height) above the burner tip. For magnesium and silver, the initial rise in absorbance is a consequence of a longer exposure to heat, which leads to a greater concentration of atoms in the radiation path. However, the absorbance for magnesium reaches a maximum at about the center of the flame and then falls off with height as oxidation to magnesium oxide takes place (the effect is not seen with silver because this element is more resistant to oxidation). For chromium, which forms very stable oxides, maximum absorbance lies immediately above the burner tip.

Several combinations of fuel and oxidant can be employed in FAAS methods. Table 4.1 shows the typical temperatures reached with most common flames together with their burning velocities. The burning velocities are critical flame parameters, as flames achieve stability only within a certain region of gas flow. If gas-rise velocities do no exceed the particular flame-burning velocity, the flame will propagate inside the burner, resulting in a flashback condition (an eventual explosion is most likely). As the gas flow is increased,

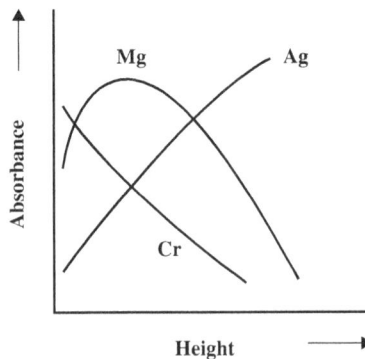

Figure 4.6. Illustrative flame absorbance profiles for magnesium, silver, and chromium, showing dependency on height of the observation of the atoms in the flame.

Table 4.1 Characteristics of some analytical flames

Fuel	Oxidant	Temperature (°C)	Maximum burning velocity (cm/s)
Acetylene	Air	2,100–2,400	158–266
Acetylene	Nitrous oxide	2,600–2,900	285
Acetylene	Oxygen	3,050–3,150	1,100–2,480
Hydrogen	Air	2,000–2,100	300–440
Hydrogen	Oxygen	2,550–2,700	900–1,400
Propane	Air	1,700–1,900	39–43
Propane	Oxygen	2,700–2,800	370–390

the rise velocity soon exceeds the burning velocity at all points and the flame rises until it reaches a steady-state point above the burner where the rise and burning velocities are just equal; this is the region for a stable flame. At higher flow rates, the flame would finally blow off. Flames that use air as oxidant can be premixed because of its relatively low burning velocity. The nitrous oxide–acetylene flame can be premixed with some care. However, more uncommon mixtures such as oxygen and acetylene require care to burn safely after premixing. Of course, oxygen and hydrogen may also form explosive mixtures, but if diluted with argon or helium, they can be premixed.

A venting system is required to remove the combustion fumes and vapors from the flame atomic absorption instrument. This venting system will protect laboratory personnel from toxic vapors that may be produced by some samples. It will help to protect the instrument from corrosive vapors which may originate from the sample, and it will remove dissipated heat that is produced by the flame or furnace.

Currently the most widely used gas combinations for FAAS are air–acetylene and nitrous oxide–acetylene. This latter high-temperature flame is very useful for those elements that tend to form heat-stable oxides in the air–acetylene flame (the "refractory elements"). However, it is not always required and may even be detrimental for analysis of many elements, because its higher temperature will cause undesired ionization of some gaseous atoms. Because of the somewhat higher burning velocity, for the hot nitrous oxide/acetylene flame, specially designed burner heads with the narrow slot are required to prevent flashback of the flame.

Figure 4.7 shows a periodic table indicating which of the two flames are recommended for each particular element. Given below are summarized

Legend:

670.8	— Wavelength (nm)
Li	— Element
0.7	— Spectral bandwidth (nm)

□ Air/acetylene
□ Nitrous oxide/acetylene

Periodic Table (values shown as wavelength (nm) / spectral bandwidth (nm)):

1	2	3	4	5	6	7	8	9	10	11	12	13	14	15	16	17	18
H																	He
Li 670.8 / 0.7	Be 234.9 / 0.7											B 249.7 / 0.7	C	N	O	F	Ne
Na 589.0 / 0.2	Mg 285.2 / 0.7											Al 309.3 / 0.7	Si 251.6 / 0.2	P 213.6 / 0.2	S	Cl	Ar
K 766.5 / 0.7	Ca 422.7 / 0.7	Sc 391.2 / 0.2	Ti 364.3 / 0.2	V 318.4 / 0.7	Cr 357.9 / 0.7	Mn 279.5 / 0.2	Fe 248.3 / 0.2	Co 240.7 / 0.2	Ni 232.0 / 0.2	Cu 324.8 / 0.7	Zn 213.9 / 0.7	Ga 287.4 / 0.7	Ge 265.1 / 0.2	As 193.7 / 0.7	Se 196.0 / 2.0	Br	Kr
Rb 780.0 / 0.7	Sr 460.7 / 0.2	Y 410.2 / 0.2	Zr 360.1 / 0.2	Nb 334.4 / 0.2	Mo 313.3 / 0.7	Tc 261.4 / 0.2	Ru 349.9 / 0.2	Rh 343.5 / 0.2	Pd 244.8 / 0.2	Ag 328.1 / 0.7	Cd 228.8 / 0.7	In 303.9 / 0.7	Sn 286.3 / 0.7	Sb 217.6 / 0.2	Te 214.3 / 0.2	I	Xe
Cs 852.1 / 0.7	Ba 553.6 / 0.2	La 550.0 / 0.2	Hf 286.6 / 0.2	Ta 271.5 / 0.2	W 255.1 / 0.2	Re 346.0 / 0.2	Os 290.9 / 0.2	Ir 264.0 / 0.2	Pt 265.9 / 0.7	Au 242.8 / 0.7	Hg 253.7 / 0.7	Tl 276.8 / 0.7	Pb 283.3 / 0.7	Bi 223.1 / 0.2	Po	At	Rn
Fr	Ra	Ac															

Lanthanides:

Ce	Pr 495.1 / 0.2	Nd 492.4 / 0.2	Pm	Sm 429.7 / 0.2	Eu 459.4 / 0.2	Gd 368.4 / 0.2	Tb 432.6 / 0.2	Dy 404.6 / 0.2	Ho 410.4 / 0.2	Er 400.8 / 0.2	Tm 371.8 / 0.2	Yb 398.8 / 0.2	Lu 336.0 / 0.7

Actinides:

Th	Pa	U 351.5 / 0.2	Np	Pu	Am	Cm	Bk	Cf	Es	Fm	Md	No	Lw

Figure 4.7.   Periodic Table with elements that can be analyzed by flame atomic absorption spectrometry.

characteristics, peculiarities, and cautions for the three more common gases used in FAAS: air, acetylene, and nitrous oxide.

Air: Air compressors are commonly used. It is strongly recommended to have a water and oil trap or filter between the compressor and the instrument. Air compressors are generally noisy and whenever possible it is advisable to locate them at some distance from laboratory workers within a suitable ventilated area. Cylinders of compressed air can be also used. However, they are recommended only as a short-term solution: a premix burner–nebulizer system uses about 20 L/min of air and, therefore, a cylinder will typically last only a few hours.

Acetylene: Suitable acetylene has typically a minimum purity specification of 99.6%. Acetylene is normally supplied dissolved in acetone, and a small amount of acetone carryover with the acetylene is acceptable (for better conditions, it may be desirable to have an acetylene filter between the acetylene tank and the instrument gas control system to remove particulates and acetone droplets from acetylene, protecting the gas controls). However, as the tank pressure falls, the relative amount of acetone entering the gas stream increases and can give rise to erratic results, as well as cause damage to instrument connecting pipes. For this reason, acetylene tanks should be replaced when the cylinder pressure drops to about 600 kPa. Furthermore, acetylene tanks should always be stored and operated in a vertical position to prevent liquid acetone from reaching the cylinder valve. As an important precaution, acetylene line pressure from the cylinder to the instrument should never be allowed to exceed 100 kPa; at higher pressures, acetylene can spontaneously decompose or even explode.

Nitrous oxide: The use of cylinders of nitrous oxide (99.0% minimum purity) requires some accessories and precautions. Nitrous oxide is supplied in the liquid state, so the pressure gauge does not give a true indication of how much nitrous oxide remains in the cylinder until the pressure starts to fall rapidly as the residual gas is drawn off. On the other hand, when nitrous oxide is rapidly removed from the cylinder, the expanding gas causes cooling of the cylinder pressure regulator and the regulator diaphragm sometimes freezes; this can create erratic flame conditions or even a flashback. It is therefore advisable to heat the regulator. All lines carrying nitrous oxide should be free of grease, oil, or other organic material, as spontaneous combustion could eventually take place.

4.2.3 Special Sampling Techniques

One of the ever-existing aims in AAS has been the improvement of the achievable DLs, which are primarily limited by the usual sample-introduction system (called "the Achille's heel" of analytical atomic spectrometry). As explained above, in the conventional nebulizer/burner systems, only about 10% of the aspirated sample solution reaches the flame, while the other 90% goes as waste. Furthermore, the transport of the sample to the flame is limited by the aspiration rate of the nebulizer. These two disadvantages were addressed already in the late 1960s and early 1970s with different approaches, including the so-called *sampling boat technique* and the *Delves cup*. Recently, such initial approaches have been displaced by the Electrothermal atomic absorption spectrometry (ETAAS) technique (see Chapter 5). They will be briefly described below to understand this problem and modern solutions to overcome it.

In the "sampling-boat technique," a small amount of the sample is introduced into a tantalum boat, the solution is dried (e.g., near the flame of an AAS or in a furnace), and then the boat is introduced into the flame of the spectrometer. The sample in the boat in the flame is rapidly and quantitatively atomized giving rise to a transient signal. The temperature that can be achieved with this technique is only around 1,200°C, but this temperature is suitable for some elements of toxicological and environmental interest, such as As, Cd, Pd, Hg, and Se.

A little later, H.T. Delves proposed a modification of the boat technique, giving rise to the so-called Delves cup, which enjoyed wide use in the determination of lead and cadmium in biological fluids. In this case, the long tantalum boat was replaced with a small round nickel cup, and an open tube was mounted over this cup to increase the sensitivity. The radiation beam passed through the tube and the analyte element was atomized into the tube. This tube increased the residence period of the atoms in the optical path.

4.3 FLAME ATOMIC ABSORPTION INSTRUMENTATION

At least 10 well-known commercial companies manufacture atomic absorption spectrometers (see Appendix A). Most of these companies have several models of AAS instruments available for customer choice. In some cases, marketed AAS spectrometers can be used both with a flame and with an electrothermal atomizer, which are easily exchanged if needed (there are also instruments available in the marketplace offering true simultaneous operation of flame and graphite furnace, with both atomizers permanently mounted and aligned for immediate use).

AAS manufacturers market an important variety of accessories for automated measurements or to enhance performance and increase the number of applications of the commercially available AAS instrumentation.

4.3.1 Flame Atomic Absorption Spectrometers

There is today a wide choice of instruments for FAAS measurements, including single-beam or double-beam optics. For background correction, the deuterium lamp (although other choices are available) is preferred in FAAS work. Some of the FAAS instruments have modular parts that can be easily replaced, thereby providing a high degree of flexibility. In some instances, they have complex and expensive optics for maximum performance, while in others more attention is paid to other practical requirements (e.g., to simplicity in order to make an AAS instrument portable for field applications).

Although in Chapter 2 it was explained why AAS is considered a "single-element technique" (implying that aspiration of every sample has to be repeated for each element to be determined), now, fast-sequential AAS instruments combined with an automatic turret for up to eight hollow cathode lamps are commercially available, enabling several elements to be determined in rapid sequence in just a "single aspiration" of each sample (this capability also allows for internal standard corrections to take place in order to compensate long-term drift or sample preparation errors). In addition, it should be stressed here that a company markets an AAS instrument with a continuous source (a xenon short-arc lamp) and a high-resolution echelle spectrometer with a UV-sensitive charge-coupled device (CCD) detector (Resano and García-Ruiz 2011). Four major advantages of the configuration are:

1. Improved signal-to-noise ratio because of the high intensity of the radiation source. Furthermore, the atomic absorption can not only be measured at the center of the absorption line (with maximum sensitivity) but also within its wings (with reduced sensitivity), thus greatly increasing the dynamic range.
2. The entire spectral environment around the analytical line then becomes measurable, providing more information than available for common AAS instruments, which eventually results in achieving a more reliable and accurate background correction.
3. Elements for which a reliable line radiation source is not available can be determined. The extremely high resolution power of the double monochromator, which provides intense and very narrow emission lines at any wavelength, makes it possible to select extremely narrow wavelengths (a few picometers wide) overlapping exactly with the molecular hyperfine structured (rotational) absorption line of the analyte-containing molecule and hence to eliminate spectral interferences. This makes possible the determination of low levels of non-metals, such as Cl, F, P Br, and S.
4. A truly simultaneous multi-element AAS measurement is possible by replacement of the one-dimensional array detector by a two-dimensional multi-array detector (as is a common practice in optical emission spectrometry).

Finally, it has to be noted that in the last few years particular attention has been paid by most companies to ensure flame safety. This has been achieved by automatically monitoring flame ignition, pressure in gas supplies and in the mixing chamber, drain status, automatic recognition of installed burner head and hence ensuring proper readjustment of the correct gas settings for the type of flame used, and so forth.

4.3.2 Accessories

Several commercially available accessories for flame atomic absorption spectrometers are marketed with slight modifications by several manufacturers, although special specific ones can be marketed by a single manufacturer. If the reader is interested, this particular information can be found through a quick search in the online webpages of the AAS companies given in Appendix A.

Further information about commercial versions of hydride or cold vapor generation can be found in Chapter 6, while information on accessories to carry out AAS measurements in flow systems is provided in Chapter 7.

4.3.2.1 Autosamplers

Many instrument manufacturers market computer-controlled fully automated autosamplers with XYZ arm movement and full random sample access. They usually have a capacity for more than 100 samples and several standards (in some designs unlimited sample capacity is claimed, since in such cases it is possible to exchange sample racks during analysis). Other features incorporated in some available autosamplers are for example: variable probe-arm speed settings for samples of different viscosities, elimination of cross-contamination by the use of a pump providing a continuous stream of clean rise solution, variable online dilution, and automated standard additions.

4.3.2.2 Atom Concentrator Tube or Slotted Tube Atom Trap

Several companies market a side-open quartz tube, with two slots along its length, which is positioned in the air–acetylene flame. The flame enters the tube through one of the slots and exits through the other and both ends of the tube. It is claimed that this system enhances the sensitivity by –two to three times when compared to that of a normal FAAS setup.

When a solution is aspirated, the atoms pass into the tube. The tube wall prevents the atoms from diffusing out of the optical path. This increases the residence time of the atoms in the light beam, thus increasing the absorbance signal. The tube also stops air diffusing into the flame, preventing formation of oxides.

4.3.2.3 High-Solid Analyzer

The analysis of samples with high dissolved solid content has always been a problem in FAAS because of the tendency of dissolved salts to crystallize and cause blockages in the nebulizer tip or in the burner. A common way to avoid this problem is to dilute the sample until the concentration of dissolved solids is less than 2% before the final analysis and, in fact, several manufacturers market automatic sample dilution systems. A high-solid FAAS analyzer is marketed allowing the analysis of solutions containing as much as 30% total dissolved solids without blockage of the nebulizer or the burner. The accessory consists of a switching valve, incorporating a sample loop of a predetermined volume (100 μL), which allows the introduction of a small volume of sample into a rinse-liquid stream and passage of the sample bolus into the spray chamber with minimal diffusion. The aspiration rate of the sample is determined by the nebulizer uptake rate and, as expected, the discrete sample system introduction produces a transient absorbance signal.

4.3.2.4 Flame Microsampler

When only a small amount of sample is available, or when the burner becomes clogged by continuous aspiration of high solids for a long time (e.g., solutions of serum, seawater, ore, and metallurgical digests), a microsampler system (syringe 50–250 μL) for direct injection can be used.

4.3.2.5 Automatic Burner Rotation

One inherent problem with conventional AAS analysis is comparatively narrow dynamic ranges (so, samples with a wide analyte concentration range cannot always be directly measured in a single calibration). To overcome such problems two options are available: to resort to an automatic online dilution or to employ a system allowing for automatic burner rotation. Such a "burner rotation system" provides a monitored adjustment controlled from the computer in order to be accurately and reproducibly positioned in the light beam in order to obtain different light-path values and so optimum absorbance measurements.

4.4 ANALYTICAL PERFORMANCE CHARACTERISTICS AND INTERFERENCES

General interferences in atomic spectrometric techniques have been classified in Chapter 2. The extent of each type of interference in FAAS in particular is addressed in this section.

4.4.1 Spectral Interferences

Spectral interferences detected are usually strongly dependent on the spectral bandwidth of the monochromator. Such interferences may arise from two main sources: (i) *atomic absorbance lines due to concomitants*; such atomic spectral overlaps are rare because the emission lines of hollow cathode lamps are very narrow (in any case, they can be avoided by choosing a different absorbance line); and (ii) *molecular absorption or scattering*. Molecular absorption/scattering can be due to the following:

1. Unvaporized solvent droplets or molecular species will cause a positive interference in AAS. Fortunately, spectral interferences by matrix components usually can be avoided by optimizing the flame composition or by adding the interfering species to standards. A potential matrix interference due to such absorption occurs, for example, in the determination of barium in alkaline-earth mixtures. The barium wavelength used for AAS analyses appears in the center of a broad absorption band for molecular CaOH, thus resulting in an interference by calcium in the barium determination. The net effect is eliminated by using nitrous oxide instead of air as oxidant: The higher temperature decomposes the CaOH and eliminates the absorption band. Spectral interference due to scattering by atomization products can occur when the sample contains high concentrations of elements that form stable oxide (e.g., of Ti or W) particles. Such particles formed in the flame can cause eventual light scattering.

2. Molecular absorption or light scattering by gas combustion products is also possible. For this case, correction is easily achieved from absorbance measurements made with a blank aspirated into the flame.

3. In old instruments where the radiation source is not modulated, spectral interferences may also arise from thermal emission of concomitants and gas combustion products (see, for example, Figure 4.8), or by stray radiation.

4.4.2 Nonspectral Interferences

For interferences other than spectral, the analyte signal itself is directly affected. In this section, special attention will be paid to considering physical, chemical, and ionization interferences.

1. *Physical interferences* are caused by differences in the physical properties of the sample and calibration standards, such as viscosity, surface tension, or density of the solution (e.g., adding sulfuric acid makes the solution more viscous and decreases the absorbance, while the addition of methanol increases the absorbance signal by enhancing nebulization efficiency and thus increasing the amount of sample entering the plasma). These can be overcome by

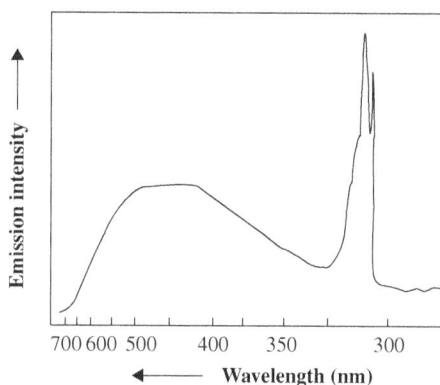

Figure 4.8. UV-Vis emission spectrum of an air–acetylene flame, showing the strong emission corresponding to the OH· bands in the 300 nm region and CH and Swan bands (C_2 molecules) between 300 and 600 nm.

matching the density and viscosity of reference solutions and samples, for example, by using the same solvent. Various authors have reported with a high degree of success by the use of sampling pumps with fixed liquid flow rates to overcome those interferences.

2. Chemical interferences are produced by the formation of a compound that prevents quantitative atomization of the analyte. There are a number of different approaches used in FAAS, which aim to overcome chemical interferences. The methods more commonly used to do so are: the selection of appropriate flame conditions, the matching of standard solutions to the sample, or the use of some special substances reducing interferences in specific chemical ways. In some circumstances, calibration by the standard addition method (see Box 2.1 in Section 2.4 of Chapter 2) allows one to overcome such interferences.

As stated in Section 2.5 of Chapter 2, chemical interferences are produced by formation of low-volatility compounds, thus decreasing the rate at which the analyte is atomized. The most typical example is the effect of phosphate on Ca absorbance (phosphate reacts with calcium ions in the flame producing calcium pyrophosphate, thus giving rise to lower absorbance signals than expected). These interferences can be reduced by adding matrix modifiers, such as "releasing agents," that is, other cations that preferentially react with the anions (e.g., Sr or La will preferentially combine with phosphate and prevent its reaction with Ca), or "protecting agents" that form volatile species with the analyte and compete with interfering species (e.g., ethylenediaminetetracetic acid [EDTA] complexes with Ca and prevents interference from phosphate and sulfate). They may be eliminated also by using a higher-temperature flame.

A serious problem occurs when the analyte reacts with flame gases forming thermodynamically stable oxides and hydroxides in the flame (e.g., aluminum, titanium, and molybdenum). Several of these elements do not exhibit appreciable absorption in the conventional air–acetylene flame. To overcome this interference the flame composition and the region of the flame monitored have to be adjusted. A more useful flame for these elements is the nitrous oxide–acetylene flame, which operates under reduction (fuel-rich) conditions. In this case, the lack of oxygen-containing species combined with the high temperature of the flame, decompose and/or prevent the formation of refractory oxides.

3. *Ionization interferences.* A noticeable fraction of alkali and alkaline-earth elements and several other elements in hot flames may be ionized in the flame, giving rise to a decreased amount of atoms available for absorbance (a decrease of sensitivity and linearity is usually observed).

$$M^0 \rightleftarrows M^+ + e^-$$

Ionization can usually by detected by noting that the calibration line has an upward curvature at higher concentrations, because a larger fraction of the atoms are ionized at lower concentrations. To overcome this interference, an "ionization suppressor," an easily ionizable element (B), for example, potassium or caesium salts, can be added, thereby providing a large amount of electrons to the flame and suppressing ionization of the analyte. Thus, if the medium contains not only M but an important amount of B species as well, and if B ionizes according to the equation:

$$B^0 \rightleftarrows B^+ + e^-$$

then the degree of ionization of M is decreased by the mass-action effect of the electrons formed from B.

4.4.3 Calibration in Flame Atomic Absorption Spectrometry

Quantitative FAAS methods are based on calibration curves, which in principle are linear. However, departures from linearity can occur and analysis should never be based on the measurement of a single standard with the assumption that Beer's law is being followed for all concentration ranges.

The most practical method of calibrating an atomic absorption spectrometer is to prepare a series of calibration standard solutions within the linear range of the apparatus response. These solutions should match as far as possible the chemical and physical properties of the samples. The absorbance values of these standards are used to plot a calibration graph. In some cases, these graphs are curved and most commercial instruments allow the choice of fit from a number of commonly used algorithms.

A widely practiced calibration method to overcome matrix interferences is the method of standard additions, which was briefly explained in Chapter 2 (see Box 2.1 in Section 2.4). The standard additions method does have some limitations, including:

1. The calibration graph must be substantially linear since accurate regression cannot be obtained from nonlinear calibration points.
2. It is essential to obtain an accurate baseline from a suitable reagent blank.
3. It may require larger quantities of sample than other methods.
4. The method is comparatively labor intensive because every sample type requires its own set of calibration solutions, and
5. Operators must be carefully trained to prepare appropriate standards (this includes recognizing if the calibration is nonlinear and diluting the standards so the respective signals lie within a linear region).

4.4.4 Analytical Figures of Merit

Table 4.2 lists some representative examples of DLs that can be achieved by FAAS (note that the LDs correspond to "instrument detection limits" or "IDLs" when a clean matrix was used. The "method detection limits" consider real-life matrices and are usually poorer than IDLs). As can be seen in Table 4.2, the achieved DLs in FAAS depend greatly on the element analyzed,

Table 4.2 Examples of detection limits (DLs) obtained by flame atomic absorption spectrometry (mg/L)

Element	DL	Element	DL	Element	DL
Al	45	K	3	Pb	15
Ca	1.5	La	3,000	Ta	1,500
Co	9	Mn	1.5	V	60
Cs	15	Mo	45	W	1,500
Cu	1.5	Nb	1,500	Zr	450

*All DLs were determined using elemental standards in dilute aqueous solutions and instrumental parameters optimized for the individual element. Data shown were determined on an "AAnalyst™ 800."

Source: Adapted with permission from "Guide to Inorganic Analysis" by Perkin Elmer.

and differences of three orders of magnitude can be expected depending on the analyte characteristics.

Under usual optimized conditions, the *precision* achievable by FAAS is of the order of 1% to 2%, but using special precautions this figure can be lowered to a few tenths of 1%.

4.4.5 Use of Organic Solvents

When organic solvents are aspirated into a flame, an oxidizing (fuel-lean) flame must be used, because the solvent must be burned. The overall atomization efficiency for producing an atomic vapor in the flame is increased by use of organic solvents in the sample solution. Such an increase is due to a variety of causes, including increased rate of aspiration, finer droplets, and more efficient evaporation and combustion of the solvent; thus, increased sensitivity is generally obtained. However, for safety reasons and except in some particular cases, the use of organic solvents in FAAS is becoming increasingly less practical.

It is obvious that much of our environment consists of water, therefore, the bulk of FAAS methods deal with water as solvent. However, the use of organic solvents for AAS is necessary for certain applications, such as for solvent extraction of metal chelates and for direct analysis of products like petroleum derivatives (a gasoline sample is too flammable to be introduced without dilution with a suitable miscible liquid in a flame instrument) or edible oils (an oil sample is too viscous to be aspirated without dilution). In such cases, particular care has to be taken with certain instrument's components because some materials used in the instruments employed for the analysis of aqueous solutions are often attacked by organic solvents.

On the other hand, there are a few solvents that should not be used in combination with FAAS, such as halogenated hydrocarbons which that readily decompose in the flame to produce phosgene (an extremely hazardous compound), very low-boiling point hydrocarbons, ethers, acetone, tetramethylfuran, and dimethylsulphoxide. Some of those solvents are so flammable that they can support a spectrometer flame without acetylene.

4.5 APPLICATIONS AND EXAMPLE CASE STUDIES

FAAS is today a well-established technique for trace metal analysis for almost any human activity area and for a wide variety of samples. Therefore, it is not easy to select a few practical and illustrative examples of FAAS applications. Many FAAS standard procedures established by international standard committees are now routinely used. Besides, "cookbooks" and application notes are provided by most companies manufacturing commercial instruments. The examples collected here will just try to give some guidelines for laboratory practice.

4.5.1 Determination of Calcium in Milk

Ca is one of the elements most frequently analyzed by FAAS. The determination of calcium in milk requires the removal of milk proteins including casein, for example, by precipitation using trichloroacetic acid. The samples are then filtered and the filtrate analyzed by FAAS (in alternative procedures the organic part of milk is removed by digestion with nitric acid and hydrogen peroxide in a microwave oven).

The slight Ca ionization that occurs in the air–acetylene flame can be controlled by the addition of an alkali salt (0.1% or more potassium as chloride) to samples and standards. However, calcium sensitivity is reduced in the presence of elements that give rise to stable oxy-salts, including Al, Si, Ti, P, V, and Zr. This effect can be reduced by the addition of 0.1–1.0% La or Sr or 1% EDTA (as a di-sodium salt).

The absorption of calcium is dependent on the fuel/air ratio and the height of the light beam above the burner. Although maximum sensitivity is obtained with a reducing (fuel-rich) flame, an oxidizing (fuel-lean) flame is frequently recommended for optimum precision. Calcium determination appears to be free of most chemical interferences in the nitrous oxide–acetylene flame. In this latter case, only ionization interferences need to be controlled, and this can be easily done by the addition of an alkali salt.

4.5.2 Determination of Molybdenum in Fertilizers

Sandy soils and those that are inherently infertile in their natural state, for example, soils low in phosphorus, are typically low in molybdenum. However, Mo is an essential micronutrient in plants (Mo is important in nitrogen metabolism and the synthesis of proteins). Therefore, fertilizers containing Mo, which can be applied to the soil or the foliage, are commercially available.

Elements forming refractory oxides in flames (such as Al, Ti, W, Cr, and Mo) require a high-temperature flame to produce adequate atomic populations, so, a nitrous oxide–acetylene flame has to be used (this flame can be dangerous if not operated properly as its burning velocity is relatively high). Unlike most of the refractory elements (with the exception of Cr, which behaves like Mo), molybdenum can be determined also in a lower-temperature air–acetylene flame; in such a case, fuel-rich conditions are required to compensate for the lower temperature in order to obtain an adequate population of Mo atoms. With the nitrous oxide–acetylene flame, a better sensitivity (about an order of magnitude improvement) can be achieved due to the higher flame temperature, more favorable flame environment, and selection of conditions for this flame is not as crucial as for the air–acetylene flame. Besides its higher sensitivity, the nitrous oxide–acetylene flame experiences much less interferences; hence, it is used almost exclusively for this determination.

In practice, interferences in the determination of molybdenum in fuel-gas-rich air–acetylene flames have been scarcely investigated and matrix matching is frequently used. Although a number of interferences may occur in a nitrous oxide–acetylene flame (e.g., calcium, strontium, iron, and sulfate), such interferences can be controlled by the addition of 0.5% aluminium chloride (aluminium seems to inhibit the lateral diffusion of molybdenum atomic vapor, leading to its increased atom concentration in the middle of the flame).

4.5.3 Determination of Lead in Gasoline

The determination of metals by FAAS in carbon-rich samples such as edible oils, motor oils, lubricating oils, and petroleum products are based on char-ashing or dilution of the sample with an organic solvent. This char-ashing technique yields accurate results for some elements while allowing the determination of trace metals at a much lower level than sample direct aspiration. The main disadvantage of this method is its tedious and lengthy procedures since the sample must be first completely carbonized on a hot plate before it is ashed in a muffle furnace. Due to this, it cannot be applied to volatile analytes that would be lost during sample preparation.

A typical FAAS determination is lead in gasoline. In this case, the method of sample preparation involves the reaction of the alkyl lead components of gasoline with iodine (the reaction with iodine reduces the variation in response for different alkyl lead compounds), followed by stabilization of the alkyl lead iodide complexes with tricaprylmethylammonium chloride (Aliquat 336) and dilution with methyl isobutyl ketone (the order of addition of the reagents must be followed explicitly: Aliquat 336 must not be added before the addition of iodine because it retards the addition of iodine and the formation of the alkyl lead iodide–Aliquat 336 complex, bringing about erroneous results).

SAFETY CAUTIONS

1. Use of FAAS with gasoline samples requires complete observation of all relevant safety practices enforced when flammable materials are analysed analyzed in a flame.
2. Anti-knock lead compounds are particularly poisonous and must not be inhaled, or ingested, or come into contact with the skin. Also, they should never be exposed to elevated temperatures (above $50°C$) or to acids and oxidizing agents.
3. Solvents used in this methodology present a hazard risk to users. Therefore, operators should consult the relevant Material Data Sheet before operation.

4.5.4 Determination of Boron, Phosphorus, and Sulfur by High-Resolution Continuum Source FAAS for Plant Analysis

During the last decade, continuous-source FAAS with a double echelle monochromator has proved useful for AAS multi-element analysis and its practical use has been established (Welz et al. 2010). It is convenient to conclude this chapter by illustrating its use in the monitoring of nutrients in crop leaves.

Foliar diagnosis is an efficient tool in agriculture to monitor the mineral nutrition of plants. The monitoring of nutrients in crop leaves allows for identifying the deficiency, sufficiency, or excess of a given element; optimization of crop production; and evaluation of fertilizer supplies. Boron, phosphorus, and sulfur are among the main nutrients usually determined in foliar diagnosis.

In routine laboratories devoted to large-scale analyses, fast sequential multi-element determination is particularly helpful because time and analytical costs may be significantly reduced. Among the spectrometric techniques available for the determination of most macro- and micronutrients in plants, conventional line-source FAAS is the oldest and most commonly used. However, the difficulty of measuring absorbance of P within the 167.167–178.765 nm and S within the 180.671–182.565 nm spectral ranges, and the poor sensitivity of non-resonance lines of P at 213.618 nm, 213.547 nm, and 214.914 nm, may be considered as the main drawbacks to the determination of phosphorus and sulfur by such an analytical technique. These drawbacks are circumvented by measuring at molecular lines of PO and CS by using a high-resolution continuous-source FAAS. On the other hand, the low sensitivity of B determination by using the line source (as a consequence of the small atomization degree of B in flames) can be improved using the high-intensity xenon short-arc lamp of the high-resolution continuous-source FAAS.

The sequential multi-element analyses, the summation of absorbance signals of main and secondary atomic lines resulting in a new calibration function, and the integration of the absorbance signal over the center pixel by including part of the line wings are also interesting facilities of this instrument to improve sensitivity (Bechlin et al. 2013).

5

Electrothermal Atomic Absorption Spectrometry

Electrothermal atomic absorption spectrometry (ETAAS) allows laboratories involved in routine analyses to achieve detection limits in the low μg/L level, and requires very low sample volumes. In addition, direct atomization from solids and slurries is also possible.

This chapter describes the crucial instrumentation components in ETAAS, such as the electrothermal atomizer and instrumental background correctors inter alia. Analytical aspects like potential interferences and how to deal with them, analytical performance characteristics, and hints for the selection of proper chemical modifiers are also given in detail within this chapter. A separate section addresses aspects related to atomization from solids and slurries. The chapter ends with three illustrative examples of applications of particular relevance in the biological, food, and industrial fields.

5.1 INTRODUCTION

As already indicated, typical detection limits obtained with conventional flame atomic absorption spectrometry (FAAS) are of the order of tenths of mg/L (ppm) or even down to the ppm level. These detection limits, however, are very high for many practical applications, such as the determination of toxic metals and metalloids in biological samples or in the environment. Three main alternatives can be considered in order to enhance the sensitivity of AAS determinations: (1) to employ an atomizer more efficient than a flame, such as electrothermal atomization in a graphite furnace; (2) to resort to analyte vapor-generation (VG) techniques (see Chapter 6); and (3) to use appropriate analyte preconcentration systems (e.g., liquid–liquid or solid–liquid), which will be dealt with in Chapter 7.

Detection limits in the low μg/L level by AAS can be obtained by flameless atomization using electrothermal atomizers. According to the International Union for Pure and Applied Chemistry (IUPAC), an electrothermal atomizer is defined as "a device which is heated, to the temperature required for analyte atomization, by the passage of electrical current through its body" (Inczédy et al. 1997). Using an electrothermal atomizer, samples may be heated inside the device for final atomization of the analyte by conductive, convective, or radiative processes.

Tubular electrothermal atomizers in which the atomizer itself defines and confines the analyte's observation volume for measurement by AAS has given rise to a very sensitive analytical technique widely used in routine laboratories; this is known as electrothermal atomic absorption spectrometry (ETAAS). The attractiveness of this atomization device relies on features such as its high vaporization and atomization efficiency, a well-controlled thermal and chemical environment, its ability to handle high dissolved solid contents, the very low volume (microliters) of sample required (offering special benefits to those cases for which the amount of sample available for analysis is limited, as in many clinical analysis), and the high sensitivity that can be achieved by ETAAS.

Since the introduction of the first commercial electrothermal atomization—atomic absorption spectrometer by the Perkin Elmer company in 1970, many instrumental improvements have been developed to reduce chemical interferences and to enhance accuracy, precision, and sensitivity. Some of the most significant instrumental advances include the use of an autosampler to improve precision, a platform or probe atomization system to approach atomization under isothermal conditions, transversely heated tubes for better isothermal heating, pyrolytically coated tubes to minimize analyte interactions with the atomizer, improved background correction methods to avoid unspecific interferences, measurement of integrated absorbance, and fast electronics to more accurately measure the absorbance signal (the electrothermal atomizer produces fast transient signals that are grossly distorted if the electronics are too slow to follow the rapid atomization phenomenon in the tube). It is also important to mention and comment on the vast amount of studies carried out searching for chemicals acting as appropriate "matrix modifiers" (see Section 5.6.).

Today, ETAAS is perceived by many to be a mature technique for ultratrace (μg/L and below) analysis at a modest cost. Others consider it to be in a period of senescence, doomed to extinction by the proliferation of more powerful, faster, and versatile (more costly, too) techniques. Whatever one's opinion is, it is fair to acknowledge that ETAAS enjoys a widespread use and many instruments are still sold every year. There are some areas where a remarkable research with ETAAS is still ongoing (e.g., solid sampling and ultratrace analysis of hydride-forming elements).

Finally, although it is not in the focus of this book, it has to be pointed out that electrothermal devices are also frequently employed for atomization in emission and fluorescence analytical atomic spectrometry, as well as for vaporization in combination with other atomization/excitation or ionization sources ("combined" or "tandem sources"). In addition, the unique physicochemical microenvironment that can be attained within the electrothermal device has been used to advantage in the domain of fundamental atomic spectrometry and physical chemistry and examples include the determination of gas- and solid-phase diffusion coefficients of high-temperature metal vapors, sublimation heat of refractory metals, and fundamental optical constants (Sturgeon 1996).

5.2 THE ELECTROTHERMAL ATOMIZER

Apart from high atomization efficiencies, a convenient electrothermal atomizer for analytical AAS measurements on a routine basis should also meet some additional criteria such as rapid formation of the absorbing atoms, sufficiently long residence time within the observation zone, complete removal of all sample constituents after each determination, rapid cycling time, stabilization of response over time, and reasonable "large" sample volume or mass capacity in order to attain good sensitivity. Critical instrumental factors to be considered for optimum ETAAS performance include the length and internal diameter of the atomization cell and the material from which it is constructed, the presence of a proper platform, the precise control of the temperature, the heating rate, and the temperature distribution during atomization.

ETAAS was first introduced by B.V. L'vov in 1959. The system proposed at that time by L'vov was designed for sample volatilization into a well-defined volume at a high and constant temperature, selected to favor complete atomization. Unfortunately, the L'vov furnace was found to be difficult to use for routine analysis, which led J.D. Massmann to introduce a simplified atomizer allowing for the direct injection of the sample onto the surface of the graphite tube (see Figure 5.1). It consists of a 5-cm-long cylindrical graphite tube that is aligned horizontally in the optical path of the spectrometer. The tube was longitudinally heated (i.e., end heated) by passing a high current at low voltage through the tube. This resistance heating permitted a fine gradation of the temperature (i.e., the temperature program could be more strictly controlled) and the atomizer was continuously fluxed with an inert gas stream (e.g., Ar) to prevent entrance of atmospheric oxygen. The design of the first commercial

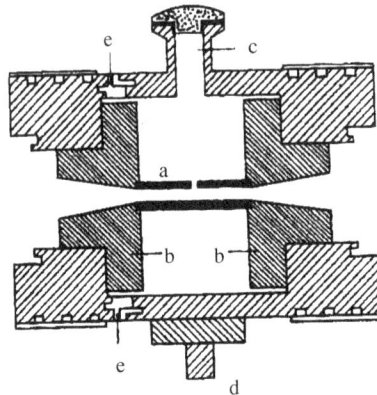

Figure 5.1. Schematic of the original Massmann design. (a) Graphite tube, (b) steel flanges, (c) sample introduction port, (d) mount, and (e) plastic insulator.

instrument was based on such a Massmann design. In the commercial instrument the tube is held in place between two graphite contact cylinders, which provide electrical connection. The heated graphite is protected from air oxidation by end windows and two streams of argon. An external gas flow surrounds the outside of the tube, and an internal gas flow entering between a window and the tube end purges the inside of the tube and exits between the other tube end and the second window. The entire assembly is mounted within an enclosed water-cooled housing.

Unfortunately, the temporal and spatial high-temperature non-isothermal behavior existing inside the tube caused problems for the analysis of real samples, in particular matrix interferences and condensation effects. It was not until 1978 when L'vov published a second, seminal paper, dealing with the concept of atomization of samples from a platform placed within the tube, by which a substantial reduction of matrix effects became possible. The efforts to achieve precise and accurate ETAAS analyses were decisively advanced with the introduction of a "stabilized temperature platform furnace" (STPF) concept of Slavin et al. upon which modern ETAAS is now based (Slavin et al. 1981). In STPF several key improvements were implemented to decrease interferences in the analysis by ETAAS. These advances include the use of an autosampler, a L'vov platform, an appropriate matrix modifier, Zeeman background correction, improved pyrolytically coated graphite tubes, rapid heating of the furnace, and fast electronics to detect the absorbance and integration of the signals.

5.2.1 The Atomization Tube

Over the years many advances have been made to the design and materials used for construction of atomization tubes (*cuvettes*) for ETAAS. Of particular interest is the reduction in size as compared with early furnaces. The smaller *cuvettes* have a much reduced thermal mass and consequently achieve operating temperatures much higher than those of the larger *cuvettes*; furthermore, atoms are confined in a smaller volume. This in turn produces a higher sensitivity and greatly reduces overall atomization times leading to extended tube lifetimes. Figure 5.2 shows a photograph and a schematic of a basic atomization tube used for analytical ETAAS. As can be seen, it consists of a cylinder (3–5 cm in length and a few millimeters in diameter) with an orifice in the center of the tube wall for sample introduction. Both ends of the tube are open to allow for the removal of sample constituents after the analysis and for passage of the light from the hollow cathode lamp for absorption measurement.

The material used for the atomizer body must be electrically conductive and be able to withstand high temperatures. In addition, good thermal shock resistance is required as well as durability and high resistance against attack by sample matrices and added chemicals. The material should also be

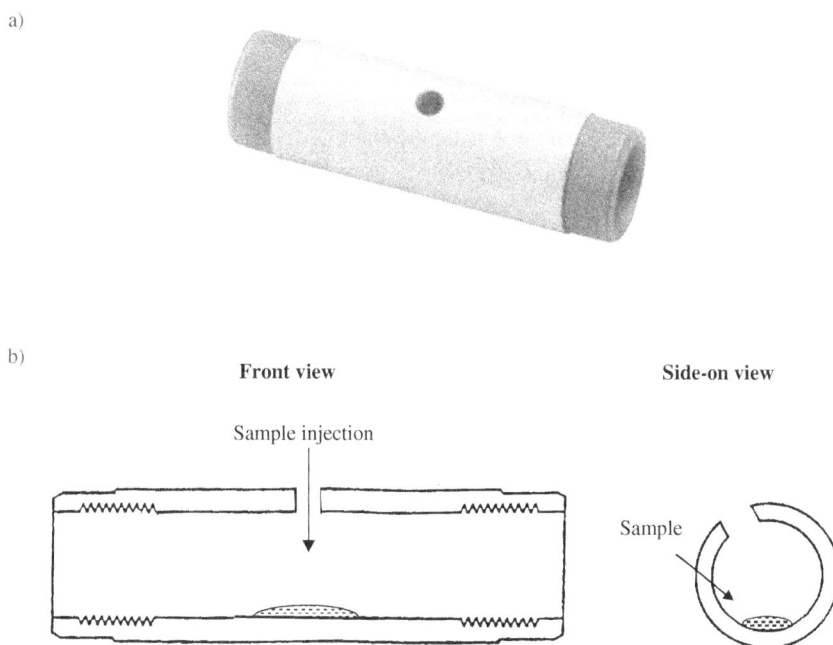

a)

b)

Figure 5.2. A basic *cuvette* for analytical electrothermal atomic absorption spectrometry. (a) Photograph. (b) Schematic views of the internal reservoir for analyte-absorbing atoms.

impermeable to gaseous sample components and to confine the atomic vapor within the atomizer for a sufficiently long time. Other requirements are high purity, low costs, and good mechanization.

In general, current electrothermal cells can be divided into non-graphitic (made from materials other than carbon, such as molybdenum and tungsten) and graphitic ones, depending on the substrate that constitutes the *cuvette*. Graphitic atomizers are most commonly used in commercial ETAAS instruments. The most frequent carbon material employed for the manufacturing of electrothermal atomizers is electrographite (EG) coated with pyrographite. EG is inexpensive and consists of fine-grained isotropic graphite (Frech 1996). Unfortunately, the material is reactive, particularly at elevated temperatures, and complex chemical reactions take place between sample constituents and EG during pretreatment and atomization steps. In addition, after subsequent analysis cycles, EG tubes lose substantial masses of graphite and their analytical lifetime is relatively short. Therefore, although EG tubes are easy to manufacture and priced competitively, their application is limited to volatile or "medium volatile" analytes that do not react with carbon. The analytical lifetime of EG tubes is drastically increased by a pyrolytical less-porous coating that provides both, a lower surface reactivity and the reduction of carbon vapor losses. Atomizers used in commercial ETAAS instruments are most commonly made of pyrolytically coated EG.

As already indicated above, an important improvement in the design of the *cuvette* used for ETAAS was the implementation of a platform (the so-called L'vov platform) to deposit the sample within the tube (see Figure 5.3). The heating characteristics of platform-equipped tubes are only slightly different from those without a platform. However, since the platform has a finite heat capacity and is heated primarily by tube radiation, its temperature will initially lag behind that of the tube wall. The sample on it will be volatilized later in time (relative to direct wall atomization), at higher tube- and gas-phase temperatures, which will favor isothermal atom formation. In other words, by using the platform the atomization time of the analyte is shifted until the tube and inert gas inside are at a more constant temperature. Thus, the platform reduces interference effects arising from the above-mentioned temporal non-isothermality typical in tube-wall atomization.

An alternative to achieving atomization under temporally isothermal conditions is to place the sample on a graphite probe. This probe is removed from the atomization tube prior to the atomization stage, and then re-introduced once the tube has reached the selected temperature. This system was commercialized for a while, but being more complex than the platform and requiring an additional aperture in the tube (reducing analytical sensitivity) its fabrication and marketing selling was discontinued.

a)

b)

Front view   **Side-on view**

Platform

Graphite tube

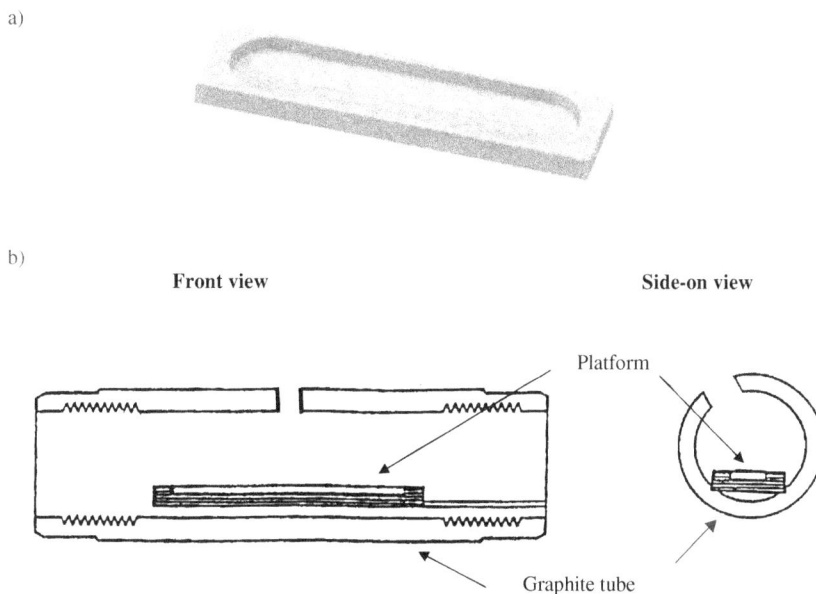

Figure 5.3. L'vov platform. (a) Photograph of the L'vov platform. As can be seen, the L'vov platform has a slight depression in the center, which can accommodate up to 50 mL of solution. (b) Schematics of the L'vov platform inserted in a graphite tube.

5.2.2 Side-Heated Atomizers

Establishing a constant temperature requires that the tube wall achieves the desired final temperature as rapidly as possible. In the conventional longitudinally heated tubes the center of the tube heats most rapidly; therefore, some time is required to establish a more or less steady temperature along the tube (i.e., to achieve isothermal conditions all along the tube at the desired atomization temperature [see schematic of Figure 5.4a]). Thus, the most significant shortcoming of the Massmann-type furnaces is the pronounced longitudinal temperature gradient that develops at high temperature (1,000 K/cm); this accounts for the majority of the matrix interferences and condensation problems encountered using the graphite furnace as an atomizer for AAS.

Trying to overcome such a fundamental limitation, a commercial transversely heated graphite atomizer was eventually introduced. In this case, a graphite tube (shown in Figure 5.4b) includes integral tabs that protrude from each side. These tabs are inserted into the electrical contacts, when power is applied, the tube is heated across its circumference (i.e., transversely). By applying power in this manner, the tube is heated evenly over its entire length (see Figure 5.4c); thus, almost isothermal heating is achieved along the tube, significantly reducing the sample condensation problems observed with longitudinally heated furnace systems.

An additional advantage of the transversely heated furnace is that it allows the use of longitudinal Zeeman-effect background correction. As described in Chapter 3 (Section 3.6.2), longitudinal Zeeman offers all of the advantages of transverse Zeeman correction without the need to include a polarizer in the optical system, thus providing an improvement in light throughput.

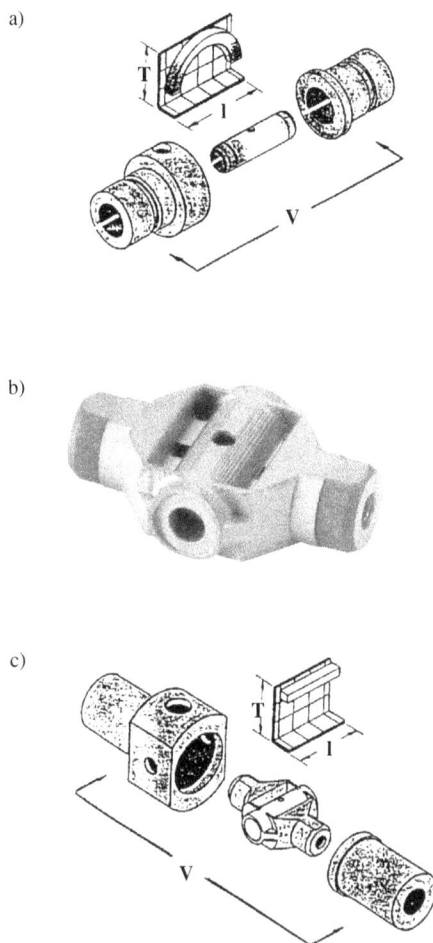

Figure 5.4. Comparison of longitudinally and transversely heated atomizers. V: voltage, l: tube length, T: temperature. (a) Scheme and heating profile of a longitudinally heated atomizer. (b) Photograph of a transversely heated atomizer. (c) Scheme and heating profile of a transversely heated atomizer.

5.3 BASIC STEPS IN ANALYSIS BY ELECTROTHERMAL ATOMIC ABSORPTION SPECTROMETRY: THE TEMPERATURE PROGRAM

A determination by ETAAS starts by dispensing a known volume of sample into the furnace. The sample is then subjected to a multistep temperature program by increasing the electric current through the atomizer body. As can be seen in Figure 5.5, these furnace-heating steps include: drying, pyrolysis, and atomization (between the pyrolysis and the atomization process a cool-down step can also be included). The furnace analysis cycle ends by a higher temperature clean-out step (when necessary). Then, the furnace cools down to enable the next ETAAS measurement. The whole process gives rise to a throughput of about 20 samples (or even less) per hour.

Important parameters to be controlled for each step are the "final temperature" during each step, the "ramp time" or time needed to achieve that temperature, and the "hold time," that is, the time maintaining that final temperature. In addition, the inert gas flow rate should be controlled through each step. We will now briefly review each of those main steps.

Sample drying: This step must be accomplished in a controlled manner, such that there is a slow and even evaporation of the solvent from the matrix. During this process sample spattering should be avoided. Temperatures around 100–120°C are commonly used for this step for aqueous dissolved samples. Of course, the use of a longer ramp time provides a "gentler" increase in heating.

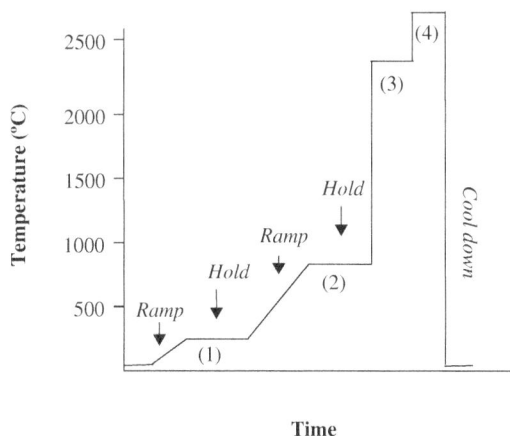

Figure 5.5. Steps of a temperature program for analysis by electrothermal atomic absorption spectrometry. (1) Drying, (2) ashing or pyrolysis, (3) atomization, and (4) clean-out step.

When a L'vov platform is used, the normal temperature lag of the platform (versus the tube wall) produces a natural ramping effect; therefore, shorter ramp times are usually required using platform atomization. The internal gas flow normally is left at its default maximum value (approximately 300 mL/min) to purge the vaporized solvent from the tube.

Pyrolysis (or Ashing): The purpose of this step is to volatilize inorganic and organic matrix components, leaving the analyte in a less-complex matrix for analysis. Therefore, the temperature is increased as high as possible to volatilize matrix components but below the temperature at which analyte losses would occur. Therefore, the pyrolysis temperature and the effectiveness of the step in matrix removal are limited by the temperature at which analyte atoms are lost. The internal inert gas flow is left again in this step at approx. 300 mL/min to drive off volatilized matrix materials (for particular sample types it may be advantageous to use an oxidizing gas, such as air, to help the in-tube sample decomposition).

With side-on heated furnaces, it is frequently advantageous to cool the atomizer prior to atomization since the heating rate is a function of the temperature interval to be covered. As the temperature interval is increased, the rate of heating also increases and this also extends the isothermal zone within the tube immediately after heating.

Atomization: In this step the temperature is increased to the point at which dissociation of volatilized molecular species occurs to form analyte atoms. The atomization temperature of choice will depend on the analyte. Care should be taken to avoid the use of an excessively high atomization temperature, because analyte residence time in the tube will decrease and a loss of sensitivity will occur. Furthermore, the use of excessively high atomization temperatures can shorten the useful lifetime of the relatively expensive graphite tubes.

For appropriate atomization it is desirable to increase the temperature as quickly as possible. In this way, the wall and atmosphere inside the tube are heated much faster than the platform, thus ensuring a stabilized tube atmosphere temperature at the time of analyte volatilization. Therefore, atomization ramp times normally will be set to minimum values in order to achieve the highest heating rate. It is also desirable to reduce or to interrupt the internal gas flow during atomization in order to increase the residence time of the atomic vapor in the atomizer.

After atomization, the furnace may be heated to still higher temperatures to burn off any sample residue that may remain in the furnace (tube cleaning) as illustrated in the final plateau of Figure 5.5. A typical cleaning-stage temperature is 2,800 °C or 2,900 °C for 2 s (when refractory elements are being determined, this tube-cleaning step becomes irrelevant since the atomization temperature for those elements is very high anyway). Finally, the furnace should be cooled down to near-ambient temperature prior to the introduction of next sample.

5.4 INSTRUMENTATION

The basic components of an instrument for ETAAS analysis are the same as for flame atomic absorption spectrometry (FAAS), namely, a line source lamp (a continuous source can be used instead, as is the case in some instruments commercially available), an atomizer, a background corrector, a monochromator, a light detector, and a data acquisition and treatment system. Apart from the obvious differences in the atomizer, ETAAS instrumentation has special requirements in some of the above components (not so crucial for FAAS analyses). For instance, particular attention should be paid to the background corrector as well as to the data acquisition and treatment system. Furthermore, it has been demonstrated that the use of an autosampler is strongly recommended for ETAAS analysis in order to achieve precise-enough analytical results. Finally, it should be mentioned that the optical system must allow maximum lamp light throughput while avoiding the incandescent continuum light generated from the inner surfaces of the heated tube.

5.4.1 Sample-Introduction System

Just a few microliters of the sample are introduced into the graphite tube for an ETAAS analysis. While skilled operators may obtain reasonable reproducibility by manual injection with a pipette, it is now proven that autosamplers provide superior results in terms of reproducibility. Autosamplers are also used to add appropriate chemicals (e.g., "chemical modifiers," which will be described in Section 5.6).

Another advantage brought about by the use of autosamplers is that the instrument can be left working unattended. This advantage is especially crucial when considering that the sample throughput in ETAAS analysis is much lower than that attained in flame AAS and, hence, the use of the autosampler may critically reduce the costs of analyses.

5.4.2 Instrumental Background Correction

Instrumental methods for background correction have been already described in Chapter 3 (Section 3.6). However, it has to be highlighted here that background absorption is the most severe problem when using electrothermal atomizers. During the atomization step any organic material still present in the atomization tube is pyrolyzed and the resulting smoke may cause severe attenuation of the light beam. The presence of salts in the atomization tube can give rise to a large background absorbance when atomized at high temperatures. All these absorptions occur outside of the atomic line and are known as "unspecific" absorption or background. In this context, the successful development of efficient methods of background correction has been a crucial aspect for the development of a wide variety of applications and the increasing importance of ETAAS. In particular, the implementation of the Zeeman's effect background correction, allowing for correction of high sample background and structured background absorption, has been a most significant advance in ETAAS analysis.

5.4.3 Data Acquisition and Treatment

Instrument electronics must be able to respond accurately to the fast, transient signals generated with ETAAS: As Figure 5.6 shows, as atomization begins, analyte atoms are formed, and the signal increases (reflecting the increasing population in the furnace), the signal will continue to increase until the rate of atom generation becomes less than the rate of atom diffusion out of the furnace. Then, the falling atom population results in a signal that decreases until all atoms are lost. Factors affecting the peak-shape characteristics are: temperature rise time of the atomization system, analyte volatility, atomization temperature, atomization tube material, and reactivity of the analyte with graphite or with the gas.

The obtained "absorbance peaks" generally appear asymmetric in that they have a steep leading edge (representing the rapid atomization of analyte atoms) and a less steep trailing edge resulting from diffusion of atoms from the atomizer; Figure 5.6 illustrates a typical peak profile. As can be seen, it approximates to a combination of two different Gaussian curves, one defining the leading edge and the other the trailing edge.

To determine the analyte content of the sample, the resulting signal (peak height or peak area) must be measured. Therefore, in order to obtain good analytical results the transient absorbance signal must be measured accurately and, so, one of the potential limitations is the inadequate speed of the instrument's electronics. For many years, measuring the height of the peak was the only practical

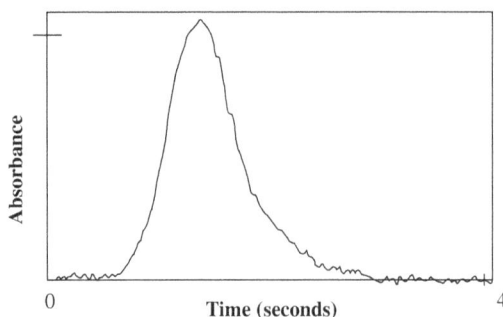

Figure 5.6. A typical absorbance peak versus time obtained with electrothermal atomic absorption spectrometry.

means to quantify the obtained transient signal. However, peak height measurements provide accurate signals for analyte concentrations only when: (1) there is a constant degree of atomization in the cell, (2) the average residence time of atoms in the cell is constant, and (3) the rate of atom formation is constant.

As matrix components can seriously affect the rate of analyte atom formation, peak height measurements are seldom used today. In fact, today's instrumentation relies on the measurement of peak areas: In principle, if the temperature in the atomizer is constant during the atomization step, the peak area will represent a count of all atoms present in the sample aliquot, regardless of the atomization rate. Peak area measurements, however, are very sensitive to variations produced in the long peak tails found with some analyses (see, for example, Figure 5.6). This will degrade the precision because very small changes toward the end of a peak tail will become significant. It has been found that a better compromise can be to use a specific peak area mode, which utilizes a short peak search time so that not all of the peak tail is included in the final measurement (analytical signal).

5.5 INTERFERENCES

In electrothermal atomization it is convenient to classify potential interferences into two broad groups: (i) spectral interferences, produced by continuous emission or by light absorption either by atoms different from the analyte or, most commonly, by molecules and smoke; and (ii) nonspectral interferences, which are those affecting the production (or the availability) of analyte-absorbing atoms.

5.5.1 Spectral Interferences

Continuous emission: Emission interference arises when the intense light emitted by the hot graphite tube or platform ("blackbody" radiation) is incident on the instrument's light detector. The magnitude of this interference depends on the wavelength (visible lines are more affected than UV lines). Emission interference is controlled basically through spectrometer optical design (in particular the slit height), in such a way that the light reaching the detector comes from the center of the atomizer and not from the tube wall or platform, which are the sources of black-body emission.

Background absorption: Background absorption is a nonspecific attenuation of light at the analyte wavelength at the atomization step. It is caused by molecular absorption (broad band) or by light scattering caused by undissociated matrix components within the light path. Therefore, background absorption derives from an insufficient pyrolysis step.

When high background absorption is limiting the quality of the analysis, it is frequently desirable to reduce the sample size: This gives rise to a reduced mass of the background-producing matrix components with a corresponding reduction in background absorption (obviously this strategy gives rise also to a lower analytical sensitivity). As the degree of background absorption is wavelength dependent, another alternative to minimize background absorption consists of using, when available, another absorption wavelength (background absorption is usually greater at lower wavelengths). Using a procedure commonly known as "matrix modification," the relative volatilities of the matrix and analyte can frequently be controlled. The procedure consists of adding a reagent ("chemical modifier") to generate either increased matrix volatility or decreased analyte volatility (see Section 5.6).

However, only in rare occasions is the use of the above strategies efficient enough in completely eliminating background absorption. Therefore, it is most common practice to compensate for residual background by resorting to instrumental approaches, described already in Chapter 3 (Section 3.6) and in Section 5.4.2 of this chapter.

5.5.2 Nonspectral Interferences

Nonspectral interferences in ETAAS have been widely studied. They can arise from several different processes: (1) gas-phase formation of analyte species, which can diffuse from the optical path, resulting in loss prior to dissociation into atoms; (2) condensed-phase reaction to form an analyte compound that is volatilized and diffused out of the absorption tube prior to atomization; (3) co-volatilization or thermal expulsion of the analyte (in a solid, liquid, or vapor phase) together with rapidly expanding matrix gases or by a carrier (or occlusion) mechanism prior to (or after) reaching the atomization temperature; and (4) changes in the atomization path modifying analyte populations.

Strategies to minimize nonspectral interferences include the use of the L'vov platform, "fast heating" in the atomization step, impermeable pyrolytic-coated graphite to reduce the tendency toward carbide formation, appropriate chemical modifiers, and also the use of the standard addition method.

5.6 CHEMICAL MODIFIERS

If the pyrolysis step were 100% efficient, that is, if all of the matrix could be driven off during this step, there would be no background absorption since the sample components that cause background absorption would be removed prior to atomization. Analyte atoms, however, must not be lost during such pyrolysis. Therefore, the pyrolysis temperature (and the effectiveness of the pyrolysis step in matrix removal) is critical and limited by the temperature at which analyte atoms are lost (Nóbrega et al. 2004).

Chemical modification, originally introduced using the term "matrix modification," has evolved into a common methodological approach in ETAAS measurements. According to IUPAC's recommendations (Inczédy et al. 1997): "In order to influence processes taking place in the atomizer in the desired way, reagents called chemical modifiers may be added. These can help to retain the analyte to higher temperatures during pyrolysis, to remove unwanted concomitants or improve atomization in other way."

This added chemical modifier allows for, generally, a thermal stabilization of the analyte (of special interest for highly volatile analytes or when various analyte species of different volatility, for example, inorganic and methylated arsenic are present in the sample), and also an efficient matrix removal by volatilization during the pyrolysis stage.

A chemical modifier is typically a concentrated solution containing one or more compounds that are added to the sample aliquot in the furnace via sequential deposition using an autosampler or directly added to a solution used to dilute the sample prior to its introduction into the tube. Many different modifiers have been employed in ETAAS. The general demands of chemical modifiers include the absence of any negative influence of the modifier on the lifetime of the graphite tube, availability in high purity, low toxicity, robustness of the modifier action, and so forth. Clearly, compounds of the elements routinely determined with ETAAS (e.g., Pb, Cd, Se, As) are not desirable as modifiers. The different modifiers can be classified according to their chemical nature into several groups: (1) inorganic salts, based on ions such as Ni^{2+}, Pd^{2+}, NO_3^-, PO_4^{3-}; (2) organometallic and complex compounds such as Ni(II)sulfonate and La(III)acetylacetonate; (3) acids, for example, HNO_3, H_3PO_4; (4) bases such as NaOH; (5) oxidants such as HNO_3 and MnO_4^-; (6) reductants such as ascorbic acid; (7) organic additives (e.g., EDTA, thiourea); and (8) gases, such as O_2 and H_2.

At present, compounds of the platinum-group metals (PGMs) appear to be the most universal chemical modifiers. Combinations of PGMs with other modifiers such as $Mg(NO_3)_2$ and organic compounds (e.g., ascorbic acid, citric acid, thiourea) are also popular. Refractory carbide-forming elements of the IVa–VIa groups of the periodic system (e.g., TaC, ZrC, NbC) are also frequently used (Ortner et al. 2002).

A widely pursued endeavor has been to develop a permanent chemical modifier (tube coating). Permanent chemical modification offers clear advantages such as an extended atomizer lifetime, lower reagent blanks, and the elimination of the step needed to add the modifier to each sample aliquot. The elements most commonly used for this purpose are metals with high melting points (such as iridium, rhodium, or ruthenium) or mixture of these elements with those forming very stable carbides (e.g., W, Ta, Zr). Tube impregnation is the easiest way to apply permanent modifiers to ETAAS tubes and platforms. Even more effective than impregnation is the deposition of permanent modifiers by electrolysis.

Action mechanisms of modifiers in ETAAS are a topic of significant scientific and practical interest and the literature contains different and often contradictory proposals for mechanisms of action of modifiers and coatings. The advantageous use of PGM modifiers seems to be based on the formation of intercalation compounds and a subsequent activation of these intercalated metal atoms. Such activated metal atoms are proposed to form strong covalent bonds with easily volatile analyte elements, which leads to their stabilization at apparently remarkably high temperatures. In the case of the use of refractory carbide-forming elements as modifiers it has been observed that analytes with a tendency to form oxoanions are especially stabilized by group IVa–VIa modifiers. This can be explained by the formation of stable bronzes (mixed metal oxides with refractory metal oxides) or heteropoly-compounds acting as stabilizing analytes, which form oxoanions on the surface of carbide-forming modifiers.

5.7 ATOMIZATION FROM SOLIDS AND SLURRIES

Traditionally, the determination of trace metals in solids by AAS has been carried out after a previous dissolution step. However, the direct analysis of the solid samples as they are received by a laboratory offers intrinsic advantages over conventional procedures, such as less risks of sample contamination or analyte losses, use of hazardous chemicals or pretreatment procedures are not required, less operational work is required, and so forth. However, there are also potential shortcomings, which will also be dealt with in this section, and need to be taken into account (see summary in Figure 5.7).

Unlike nebulization techniques, ETAAS does not suffer significantly from particle size effects because it offers a longer residence time and the availability of commercial instrumentation supporting solids analysis with the graphite furnace; for example, accessory developments (e.g., SSA 600 and SSA 6 from Analytik Jena AG) have led to its successful application for the analysis of a wide range of materials. When compared with other techniques for solids analysis, ETAAS offers high sensitivity and simplicity at moderate cost.

Solid sampling was first employed with a commercial atomizer in 1971. Unfortunately, some problems were associated with the direct solid sampling technique. For example, the small sample sizes required frequently were not

ATOMIZATION FROM SOLIDS AND SLURRIES
WITH ETAAS

ADVANTAGES

- No sample pretreatment.
- Low risk of sample contamination.
- Low operational work.
- Work with hazardous chemicals is minimized.
- No analyte loss.
- Low sample amount
- High sensitivity.
- High sample throughput.

DISADVANTAGES
- Difficulties in calibration.
- Low precision.

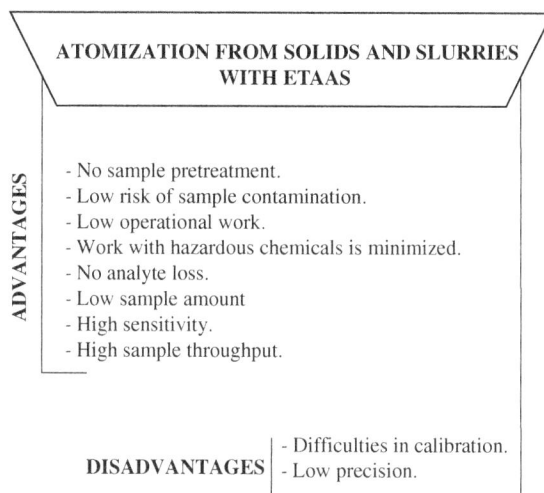

Figure 5.7. Potential advantages and shortcomings of electrothermal atomic absorption spectrometry methods for atomization from solids and slurries.

representative enough of the sample and also matrix-matched standards were required for many applications. Many of these inconveniences can be minimized by slurry sampling. Slurry sampling was introduced in 1974 and its use steadily grew until the 1990s, when it surpassed direct solid sampling. Today, slurry sampling-ETAAS can be considered as a mature technique, widely utilized for metal determination in both organic and inorganic matrices, even for routine analysis (Cal-Prieto et al. 2002). Due to the advantages of direct solid sampling (e.g., best limits of detection because of the absence of any dilution and minimal risk of contamination) it should be noted that in this 21st century the use of direct solid sampling seems to be revisited, in particular for fast screening analyses (Vale et al. 2006).

It can be considered that the slurry sampling approach combines the advantages of both the direct solid and liquid sampling methods: (1) while solid sampling requires one to weigh a very low mass of sample for each replicate and hence contamination risks during sample handling can be significant, slurry sampling allows for obtaining several replicates from just one slurry and higher amounts of sample can be weighed; (2) materials with high analyte content can be easily diluted using slurries whereas the direct solid analysis requires the addition of graphite powder, which can increase the blank values; (3) a more effective chemical modification can be carried out with slurry sampling; (4) aqueous calibration and the standard addition method have been applied successfully for quantitative analysis, which are much more troublesome for direct solid sampling; and (5) the conventional atomizers and injection systems used for liquid samples can be employed for slurries.

Even after using slurry sampling some limitations remain. Probably the most critical factor is the need for maintaining the stability of the slurry until sample injection. Particle size may seriously affect the reproducibility and accuracy if it is nonhomogeneous or too large. On the other hand, owing to the absence of a sample pretreatment step to eliminate the matrix, the molecular absorption signal is often rather high and structured making necessary the use of powerful background correctors. Another limitation is the likely buildup of carbonaceous residues in the atomizer (which reduce the lifetime of the atomizer). To minimize such inconveniences several experimental parameters have to be controlled for an optimum slurry analysis:

1. *Slurry diluent*: 0.5%–5% nitric acid is the most common diluent used for slurry preparation. It operates as an oxidizing agent and it can be also considered as a chemical modifier. It can efficiently increase the analyte extraction from the particles, increasing the suspension stability and improving precision.

2. *Stabilizing agent*: These are used to wet solid materials dispersing agglomerates and/or to avoid particles settling out. The most frequently used ones are ethanol, glycerol, and the nonionic surfactant Triton X-100.

3. *Particle size*: In general, particle size is a less critical factor for slurry analysis by ETAAS than for other atomic techniques where nebulization is required and the residence time for the solid particles in the atomizer is shorter (e.g., inductively coupled plasma-optical emission spectrometry [ICP-OES]). In ETAAS the optimum adequate particle size for a particular situation depends on many factors including sample homogeneity and density, required accuracy and precision of the results, and so forth. While good results have been reported for particles as large as 300 and 150 μm in diameter for relatively soft materials such as biological tissues, in general, major benefits are achieved with less than 50 μm diameter particles in the final slurries introduced in the tube.

4. *The mass-to-diluent-volume ratio and the analyte partitioning*: To establish an adequate mass and diluent volume for the preparation of the slurry, both the sample analyte concentration and the linear analyte response range have to be considered. Concentrated slurries show a reduction in precision because of the difficulties encountered in pipetting and higher matrix effects. On the other hand, the precision also decays when using highly diluted slurries owing to the statistical limitations of the technique when working with a small number of suspended particles.

   The percentage of analyte extracted into the slurry liquid medium also affects the achieved precision and accuracy: the higher the analyte extraction the lower the RSD. However, 100% extraction is not an essential condition for accurate analysis.

5. *Homogenization system*: A proper homogenization system of the slurry is necessary. Manual shaking, magnetic stirring, vortex mixing, gas bubbling, and ultrasonic agitation have been proposed as agitation modes to achieve slurry homogenization. In particular, the introduction of commercial systems with an ultrasonic probe allowing the direct slurry preparation and agitation in the autosampler cups constitutes a most positive advance in terms of versatility, operational simplicity, and obtaining optimum results.

5.8 ANALYTICAL PERFORMANCE CHARACTERISTICS OF ELECTROTHERMAL ATOMIC ABSORPTION SPECTROMETRIC METHODS

In ETAAS the whole sample is vaporized and all the analyte is atomized. The atoms are confined within the tube (and so within the light path that passes through the tube) for an extended period of time. As a result, sensitivity and detection limits are significantly improved as compared with FAAS, where the burner–nebulizer system acts as a relatively inefficient sampling device. In fact, ETAAS allows the elemental determination in microliter sample volumes with detection limits typically 100 to 1,000 times better than those of FAAS. Precisions achieved today, provided that sample contamination risks are carefully controlled, lie in the range of 0.5%–5%.

Tips to eliminate contamination risks

Due to the high sensitivity of this technique, one of the most common reasons for erroneous results in ETAAS is the contamination of samples, standards, or blanks. Contamination can occur at any stage in the analytical procedure and from many diverse sources. It may arise from the reagents used in sample preparation, such as any acids needed for a dissolution step, or from dirty glassware not cleaned properly after previous use. Even the laboratory atmosphere can be a problem. To avoid contamination problems, we are listing below 10 precautions that are highly advisable for routine ETAAS work:

1. A high-purity water supply, for example, deionized water having a minimum resistivity of 10 MOhms/cm, is necessary. It is advisable to produce water as and when required rather than to store it for later use.
2. Solutions of low concentration (20 ng/mL and below) should be prepared immediately before use.
3. When solutions of the same concentration of an element have to be prepared regularly it is advisable to keep the same apparatus for the same solutions.
4. All glass vessels prior to use should be washed, rinsed, and then soaked in 10%–20% nitric acid for at least 24 hours. Plastic vessels should be soaked in 1%–5% nitric acid. Finally, they should be thoroughly rinsed in high-purity water.

5. A clean bench area should be reserved for all solution preparation, as far removed as possible from sources of dust and fumes.

6. The complete ETAAS system should, preferably, be in a clean separate room.

7. The room should be under positive air pressure, supplied from an air-conditioning system with filtration against dust particles.

8. Samples must be protected from contamination by keeping sample containers sealed until they are to be processed and by minimizing the time they are open to the atmosphere.

9. For many analyses, acid dissolution is necessary for sample pretreatment. Contamination introduced from the acid used can be serious, even when using analytical-grade reagents and, therefore, high-purity reagents should be used.

10. Last but not least, sample handling does not start when the sample reaches the laboratory; special care is needed in the sample collection and in all the steps undertaken previous to its arrival at the laboratory (of course, good analyst and client communication is required here).

It has to be noted that the magnitude of the ETAAS signal observed depends on the analyte mass rather than on concentration. Therefore, it is also common to express detection limits in mass units (pg). This means that detection limits can be lowered by increasing the sample volume inserted into the graphite tube. In practice, the fact that the maximum sample size that can be accommodated in the tube is limited should be kept in mind. The maximum volume that may be pipetted onto a L'vov platform is 50 μL, but if a large concentration of nitric acid or a surfactant is added, that maximum sample volume must be reduced. Due to likely spreading of the sample using too large a sample volume may result in the spilling of the drop out of the platform.

Considering expected interferences: While spectral interferences are uncommon, nonspectral interferences are more serious and they demand appropriate instrumentation and methodology (e.g., proper graphite tube, proper background corrector) and also an appropriate chemical modifier in order to minimize their effects.

Being based on atomic absorption, ETAAS is essentially a single-element technique. ETAAS analysis times are longer than those for flame sampling, mainly due to the need to thermally program the system in order to remove solvent and matrix components prior to atomization. Therefore, ETAAS has a comparatively low sample throughput: A typical determination by ETAAS normally requires 2–4 min per element for each sample. Fortunately, as a counterpart, the ETAAS technique does not need attended operation if sample introduction is accomplished with an autosampler; thus, it can be left running all day long.

5.9 APPLICATIONS AND EXAMPLE CASE STUDIES

The high sensitivity of ETAAS and its ability to be able to analyze very small samples significantly expands the capabilities of atomic absorption spectrometry. The results produced make the technique particularly attractive for the analysis of toxic and essential ultratrace elements in clinical samples, for which a limited amount of sample is usually available. Elemental analysis of biological specimens can be carried out with good accuracy and precision following fast pretreatment within the atomizer during the pyrolysis step, thus avoiding unnecessary digestion or deproteinization procedures (and, thereby, minimizing sample manipulation and contamination risks).

ETAAS is also routinely used in many other applications where ultratrace elemental analysis is required. Food analysis, environmental control, and detection of ultratrace element impurities in high-purity materials are other important fields of potential applications. As pointed out before, the low limits of detection of the technique demand careful care in order to avoid sample contamination. For this reason it is recommended to undertake sample pretreatments and analysis in a clean room; all the reagents should be of maximum purity, containers should be especially washed, and so forth.

The determinations of a toxic metal (lead in biological fluids), an essential element (selenium) in the main food source for newborns (breast milk), and sulfur (in this case by resorting to the CS molecular band) in coal and ash slurries, have been selected and described below as typical applications of ETAAS today.

5.9.1 Determination of Lead in Human Urine and Blood

Lead is a nonessential, toxic element that has been shown to be particularly harmful to young children (in fact, lead poisoning has been described as the number one environmental disease affecting children in the United States). The recommended clinical test for assessing lead exposure is the determination of lead in whole blood and urine. Although many different approaches have been proposed for lead determination in human urine and blood by ETAAS, probably the most popular one is based on the dilution of whole blood with a phosphate modifier, followed by L'vov platform atomization and integrated absorbance measurement. Phosphate is a "hard base" (according to the Pearson classification), and tends to bind those analyte ions that are "hard" or "borderline acids" such as those of Pb, Ag, Cd, and so forth, forming precipitates that are transformed during thermal pretreatment into pyrophosphates and other intermediate species. The procedure is very simple because samples are just diluted with a solution containing the surfactant Triton X-100 (surfactants reduce the surface tension of liquid samples, providing better contact, larger surface coverage, or just better dilution/dispersive effects in the sample), nitric acid (which facilitates ashing of the sample), and the phosphate matrix modifier. In particular, $NH_4H_2PO_4$ is widely used as a modifier (Parsons et al. 1993), thus combining the beneficial effects of both phosphate and ammonium moieties (the ammonium combines with the chloride ions from the sample forming the highly volatile ammonium chloride to be eliminated during ashing).

In recent years a lot of effort has been also devoted to investigation of using permanent modifiers for lead analysis. Permanent modifiers successfully tested include Ir, Zr+Ir, W+Ir, W+Rh, Zr+Rh, or Ir+Rh. For example, for the case of iridium up to 1100 firings were possible with the same coating without sensitivity losses (Grinberg et al.2001). The samples were treated with 0.1% Triton X-100 and 0.2% nitric acid and the selected pyrolysis temperature for both matrices was 800°C.

5.9.2 Determination of Selenium in Human Milk

Selenium is an essential trace element for humans, as a component of two enzymes, glutathione peroxidase that protects against oxidative damage and deiodinase (deiodinases can either activate or inactivate thyroid hormones). It has also been identified in human plasma taking part in the selenoprotein P. Low selenium intake has been associated with the Keshan disease, a cardiomyopathy in children, and the Kashin–Beck disease in children and teenagers. Newborn babies, particularly those of low birth weight, may be at greater risk of selenium deficiency. This element seems to exhibit a rather narrow essential range because the toxic doses are only about 100 times higher than those required for normal function; the accurate and sensitive determination of selenium in foods—including, of course, human milk—is therefore of particular importance.

Selenium is considered as a volatile element and its ETAAS determination is affected by losses during ashing; so thermal stabilization by chemical modification is essential. Other difficulties include the fact that in the complex milk matrix, selenium may exist in a wide variety of chemical species. Moreover, the resonance wavelength of 196.0 nm is in the far-UV region where molecular absorption can cause serious problems. In particular, spectral interferences caused by the thermal decomposition of iron and phosphorus compounds are difficult to correct for with the deuterium lamp; therefore it becomes advisable to resort to the Zeeman background corrector when analyzing serum, urine, and so forth. Although in some reports interference-free conditions have been claimed for Se, the majority of published methods use matrix-matched standards, that is, by adding standards to a sample without detectable selenium or using a synthetic solution with the corresponding inorganic composition as the blank.

In food analysis, several metals have been used as chemical modifiers for selenium determination by ETAAS, such as nickel nitrate and palladium nitrate. Furthermore, the combination of a palladium modifier with ascorbic acid as a reducing agent has recently demonstrated its utility in the analysis of total selenium in samples such as wheat flour (Alexiu et al. 2005). Furthermore, it has been demonstrated that ammonium molybdate reduces the phosphorus spectral interference when measuring selenium by ETAAS with conventional deuterium background correction, using mussels and clams as "model samples."

For the particular case of analysis of human milk, an ETAAS method was developed based on sample digestion with a $HNO_3 + H_2O_2$ mixture in a microwave oven and determination using a mixed modifier Zr–Ir (the modifier solutions were prepared by dissolving the appropriate amounts of the salts $ZrOCl_2 \cdot 8H_2O$ and $Na_2IrCl_6 \cdot 6H_2O$). The method was applied to the analysis of breast milk of Greek women; results showed that the selenium content was in the range 16.7–42.6 µg/L (Theodorolea et al. 2005).

5.9.3 Determination of Sulfur in Coal and Ash Slurry

Sulfur cannot be directly determined by AAS because its main resonance line occurs at 180.67 nm and the two other lines of the element at 181.97 nm and 182.56 nm occur in the vacuum–UV region, a spectral region that is not accessible with conventional instrumentation. High-resolution continuum source molecular absorption spectrometry can detect the rotational and vibrational bands of diatomic molecules and allows for visualizing the entire spectral environment over several tenths of a nanometer in the vicinity of the band maximum. This strategy enables successful determination of nonmetals like sulfur, fluorine, bromine, and chlorine.

For instance, sulfur exhibits a relatively strong molecular absorption spectrum, attributed to CS, for which the most intense band of the spectrum appears around 257.59 nm. Therefore, this wavelength can be used for determining sulfur (Nakadi et al. 2013) in slurries by high-resolution continuum source molecular AAS. In other words, though this method could be classified as "non-atomic," it has been considered appropriate to describe it here as this application is an example of the present extension of the use of ETAAS (using a continuous source line) to solve practical problems.

To prepare the slurries (with a sample concentration of approximately 1 mg/mL), the samples need to be accurately weighed and suspended in diluted nitric acid solution to a known volume (surfactant Triton X-100 is also added to a final concentration of 0.04% w/v in solution in order to stabilize the slurry and to improve the CS molecule generation in the furnace). Regarding the furnace conditions, high pyrolysis temperature cannot be used because sulfur loss may occur and so pyrolysis at 300 °C is recommended.

6

Hydride Generation and Cold-Vapor Atomic Absorption Spectrometry

The generation of analyte volatile derivatives, prior to their introduction into an absorption cell, offers attractive analytical advantages that are mainly related to important sensitivity increases in the determination of many metal(loids) by atomic absorption spectrometry. In this chapter, chemical mechanisms of hydride and cold-vapor generation by the use of tetrahydroborate (III) are discussed, basic instrumentation used is described, and analytical performance characteristics are critically discussed. Approaches trying to avoid the use of tetrahydroborate (III), such as electrochemical generation, are presented as well. Also explained are common trapping/preconcentration systems for the volatilized analytes used in order to further increase sensitivity.

Three interesting and currently useful applications are described in the last section; these are determination of arsenic in water, determination of total mercury and methylmercury in hair, and, finally, the determination of selenium in bean and soil samples.

6.1 INTRODUCTION

The formation of volatile derivatives of the analyte, as a previous step for a more efficient analyte introduction into the atomizer, is now a well-established subdiscipline in atomic absorption spectrometry (AAS) in order to improve detection limits (DLs) of flame AAS (FAAS) and, more recently, also for Electrothermal atomic absorption spectrometry (ETAAS).

Vapor-generation (VG) techniques coupled to AAS detection comprise three main stages: (1) generation of the volatile analyte derivative from the liquid (or solid) sample and its transfer to a gaseous phase; (2) its collection (optional) and transfer to the atomizer by a flow of a purge gas; and (3) its decomposition in the spectrometer light path to provide the gaseous metal atoms. VG allows for a high transport efficiency of analyte into the atomic absorption cell and this is its most significant advantage in comparison with conventional flame atomic absorption spectrometry (FAAS): In typical FAAS, the nebulization process restricts the sample-introduction rate; furthermore, only a small fraction of the sample nebulized (e.g., 5%–10%) ever reaches the flame, with the remainder being directed to waste. Besides, in FAAS, the sample introduced into the flame remains in the light path for a very short period ("residence time") as it is propelled upward through the flame. However, in VG-AAS, all the volatilized analyte is transported to the measurement absorbance cell, which usually has a longer light path and also provides increased residence times in the atomizer, thus increasing analyte signals.

Varied chemical systems have been reported in analytical atomic spectrometry for VG of up to 30 elements, including generation of volatile fluorides, chlorides, carbonyls, hydrides, atoms, but only a few offer competitive performance characteristics (Wu et al. 2010). Most effective are those chemical systems based on: (1) reactions in aqueous media, (2) fast kinetics of VG and easy stripping-off of vapors from solution, (iii) reliable interference control, (iv) high and reproducible chemical yields of volatile products, and (v) volatile derivatives readily atomized in a quartz tube or graphite furnace atomizer rather than in flames (Tsalev 1999).

Tetrahydroborate (THB) salts (mainly $NaBH_4$ and KBH_4) and other borane complexes are most commonly used for volatile-species generation and subsequent determinations of trace and ultratrace levels of Ge, Sn, Pb, As, Sb, Bi, Se, Te, Hg, and Cd. More recently, several transition and noble metals have been made volatile under appropriate conditions and shown to be amenable to sensitive determinations coupled with different atomic detection techniques.

Numerous volatile and semivolatile metal compounds are present in our environment as a consequence of both anthropogenic and natural processes [Sturgeon et al. (2002)]. For example, metal hydrides can be formed in many environmental places where reducing conditions prevail and several reports have highlighted detection of metal hydrides arising from landfill and waste-water treatment sites. Formation of metal hydrides in nature is primarily considered to be the result of bacterial activity. Other major sources for the volatilization of trace elements include biotic and abiotic alkylation (usually methylation) processes. The presence of volatile molybdenum and tungsten carbonyl species has also been reported in municipal waste deposit sitesIt has been also reported the presence of volatile molybdenum and tungsten carbonyl species in municipal waste deposit sites. Also, a less known organometallic pollutant, the gasoline additive methylcyclopentadienylmanganese tricarbonyl, has also been detected in various environmental places. On the other hand, the formation and release of elemental vapors also occurs in the environment: ionic mercury can be reduced to atomic mercury and be widely dispersed.

6.2 VOLATILE HYDRIDE GENERATION BY TETRAHYDROBORATE (III) IN AQUEOUS MEDIA

Volatile hydride generation (HG) offers a practical and sensitive way to the determination of several important elements overcoming problems when determinations are attempted by straightforward nebulization AAS methods. In particular, the hydrides of group 14–16 elements are of special importance for the practical AAS determination, especially for As, Bi, Ge, Pb, Sb, Se, Sn, and Te. More recently, the scope of HG has been expanded to an increasing number of other elements, including In, Tl, Cu, Zn, and many other transition metals such as Ni, Co, Cr, Fe, or Ti, and the noble metals Au, Ag, Pd, Pt, Ir, Os, Ru, and Rh. The nature of the volatile-reaction products obtained from HG of transition and noble metals has not yet been fully characterized. HG of those latter elements has found limited analytical applications in comparison with the classical hydride-forming elements. This is mainly due to the fact that the HG efficiency for such elements is generally much lower than that for the classical hydride-forming elements (D'Ulivo et al. 2011).

The early chemical systems for volatile HG used a metal–acid reduction system, such as Zn–HCl. However, this reaction is limited to a few analytes, is slow, difficult to automate, suffers from large blanks due to reactants' metal impurities, and, in general, it is not efficient (as a result of incomplete reaction and/or entrapment of the hydride in the precipitated zinc sludge). In today's world, the almost universally used system for HG is based on sodium or potassium THB in an acidic medium. This approach allows for a fast derivatization process directly from aqueous media.

6.2.1 Mechanisms of Hydride Formation

To describe the mechanism of HG in aqueous solution with THB and ionic or molecular species of analyte elements, it was usually postulated that the active species in the reduction process is atomic hydrogen or "nascent hydrogen," which was thought to be formed during the acid hydrolysis of THB:

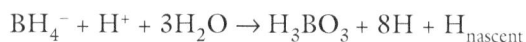

$$BH_4^- + H^+ + 3H_2O \rightarrow H_3BO_3 + 8H + H_{nascent}$$

The resultant atomic hydrogen then reacts with aqueous ions of the element M, forming the volatile hydride:

$$M^{m+} + (m + n)H \rightarrow MHn + mH^+$$

where m corresponds to the oxidation state of the analyte and n is the coordination number of the hydride. The excess unreacted atomic hydrogen forms molecular hydrogen, through reactions of the type:

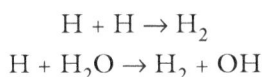

$$H + H \rightarrow H_2$$
$$H + H_2O \rightarrow H_2 + OH$$

More recently, a careful survey of the chemical literature of borane complexes together with some dedicated experiments and the use of deuterium-labeled reagents have contributed to the clarification of some controversial points concerning the mechanisms of HG, leading to the definitive rejection of the "nascent hydrogen" hypothesis, and the adoption of a reaction model based on the direct transfer of hydrogen from boron to the element through analyte–borane complex intermediates (D'Ulivo et al. 2011). For example, in studies on arsine generation using deuterated reagents and mass spectrometry, it was ascertained that the reaction of both As(III) and As(V) with $NaBD_4$ in 3M HCl and H_2O produces AsD_3 as the main product. When the reaction was performed using $NaBH_4$ and arsenic species in DCl and D_2O, the main product was AsH_3. In other words, at elevated acidities, arsine is produced by direct transfer of hydrogen to the analyte from boron. This is in contrast to the nascent hydrogen mechanism, which should give rise to all the possible randomly deuterated species $AsHnD_{3-}{}^n$ ($n = 0, 1, 2, 3$). This evidence supports a mechanism based on the direct transfer of hydrogen atoms from a borohydride reagent to arsenic and the observations are of particular relevance because they suggest that some active species containing boron–hydrogen bonds could still exist in solution, even at elevated acidities, where the rate of decomposition of THB is expected to be very fast. This observation, combined with the mechanism of hydrolysis of THB in acid solution results in the conclusion than in HG an important role is played by the BH_4^- intermediates trihydroboron, dihydroboron, and monohydroboron species, which appear to be more resistant to hydrolysis than BH_4^- itself, under strongly acidic conditions.

6.2.2 Basic Instrumentation

The chemical VG procedure can be carried out either in batches or in flow modes (Inczédy et al. 1998). In both cases, the gaseous species generated are introduced with a flow of a purge inert gas into the instrument's optical path for measurement by AAS:

1. In the *batch generation mode,* specified volumes of the sample and reactant solutions are mixed at the beginning of the measuring process inside a batch generator. A batch generator (see Figure 6.1) is a vessel made of glass or plastic serving both as the reactor as well as the separator of gases from the reaction mixture. Gas liberation in the generation vessel is accelerated by stirring, agitating, or by passing an inert purge gas (e.g., nitrogen or argon) through the solution. Released hydride is thus transported by the purge gas flow to the atomizer together with the hydrogen formed from reducing-agent decomposition. After finishing the hydride evolution, the reacted mixture is disposed to waste, generator rinsed, and a new batch or sample is then added. The analytical AAS signals recorded have a peak-shaped function. The sensitivity can be further increased by using larger sample volumes: Since all of the analyte contained in the sample is released for measurement, increasing the sample volume means that more atoms of the analyte are available to be transported to the absorbance cell. However, it has to be noted that the relationship "sample volume—sensitivity" is not linear (at high volumes the sensitivity is not as high as, in principle, expected) and this has been attributed to factors such as higher solubility of the hydride when the sample volume is high and also to hydride losses due to interaction of the hydride with a larger reaction flask.

2. In the *flow mode,* the sample, either injected as a discrete volume or continuously flowing through a flow channel, mixes with a flow of reductant. As can be

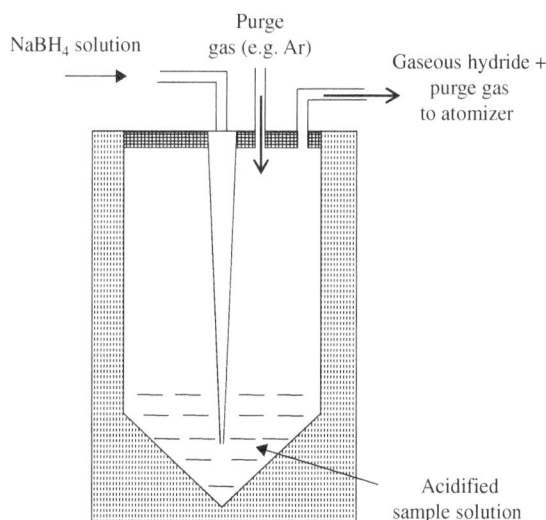

Figure 6.1. Schematic of a batch chemical vapor generation system.

seen in Figure 6.2, additional channels may be used for sample pretreatment. The mixed stream is online connected to a gas–liquid separator: The generated gaseous species are then transported to the atomizer (online connected) while the liquid goes to a waste. In the first case (flow injection mode, see Figure 6.2b) a peak-shaped signal is observed, while in the second (continuous mode, see Figure 6.2a), a steady-state signal is obtained. Advantages brought about by the use of the flow injection mode are lower sample memory effects (the carrier stream acts by cleaning the flow system and the atomizer between sample injections) and a higher sample throughput.

Flow systems for the determination of volatile derivatives of the analyte(s) are very commonly used and they are described in detail in Chapter 7.

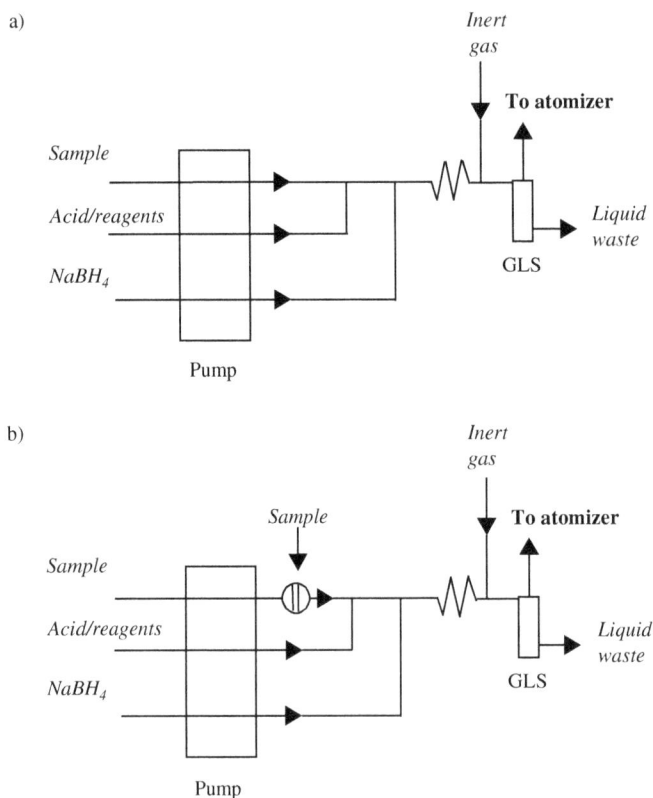

Figure 6.2. Flow configurations for hydride generation. GLS: gas–liquid separator. (a) Continuous mode. If the sample was previously offline conditioned with acid and reagents, the corresponding channel in the flow system is not needed, then only two channels become necessary: one for the conditioned sample and the other for NaBH$_4$. (b) Flow-injection mode. Usually the sample volume injected is low (e.g., 100 mL) or has been offline conditioned (acid and reagents); in such cases, the intermediate merging channel in the diagram containing acid/reagents is not needed.

Currently, the atomizers in HG-AAS used most often are heated quartz tubes and graphite furnaces located in the optical path of the atomic absorption instrument. Figure 6.3 shows a schematic diagram of a typical quartz cell, usually made of clear fused quartz tubing, with a "T" configuration (a central inlet tube for the hydride and the carrier gas, a long light path, and two gas exits). The quartz absorption cell can be heated by a flame (the cell is mounted over the burner head of the spectrometer) or by an electric cell heater. In either case, the hydride gas is dissociated in the heated cell (e.g., 800–1000°C) giving rise to free atoms of the analyte.

Finally, it is important to note that currently most manufacturers of instruments for AAS commercialize accessories allowing for both batch and flow-vapor generation.

6.2.3 Limits of Detection

Table 6.1 shows a comparison of DLs obtained for several volatile hydride-forming metals using different optical spectrometric techniques: conventional FAAS, HG coupled to AAS, ETAAS, and inductively coupled plasma-optical emission spectrometry (ICP-OES). As can be seen, DLs achieved by HG-AAS are several orders of magnitude lower than those with conventional FAAS. Also, they compare favorably with those obtained by ICP-OES and, in most cases, are in the same order as DLs obtained by ETAAS.

To account for the important improvements obtained in comparison with conventional FAAS, one has to take into account aspects pointed out in the introduction of this chapter. In HG the inefficient sample-nebulization step is not needed, the residence time of the analyte in the optical path is larger, and the optical path is longer. An additional factor accounting for the sensitivity increase in comparison with conventional FAAS is that the background is much lower since most of the analytes forming volatile hydrides have their resonant lines in the region surrounding 200 nm, where the flame produces a considerable background, when the HG procedure is carried out, the analyte is atomized in a flame-free environment (argon or nitrogen are the gases in the optical cell).

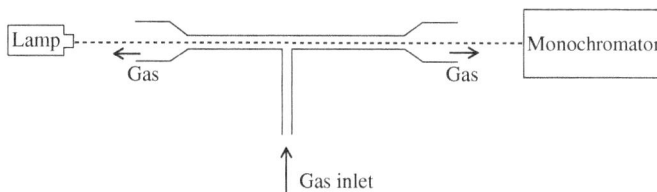

Figure 6.3. Quartz cell schematic used as an atomizer in HG-AAS.

Table 6.1. Atomic spectroscopy detection limits (DLs), in μg/L, for five selected analytes permission © by Perkin Elmer

Element	FAAS	Hydride-AAS	ETAAS	ICP-OES
As	150	0.03	0.05	2
Bi	30	0.03	0.05	1
Sb	45	0.15	0.05	2
Se	100	0.03	0.05	4
Te	30	0.03	0.1	2

- All DLs obtained by AAS were determined using instrumental parameters optimized for the individual element, including the use of electrodeless discharge lamps where available. Data shown were determined on an "AAnalyst™ 800 (Perkin Elmer Ltd.)."
- Hydride detection limits shown were determined using an "MHS-15 Mercury/Hydride system (Perkin Elmer Ltd.)."
- ETAAS detection limits were determined on an "AAnalyst 800" using 50 μL sample volumes and an integrated platform.
- All ICP-OES ("Optima 4300/5300 [Perkin Elmer Ltd.])" DLs were obtained under simultaneous multi-element conditions with the axial view of a dual-view plasma using a cyclonic spray chamber and a concentric nebulizer.

Source: Adapted with permission from "Guide to Inorganic Analysis" © by Perkin Elmer. All DLs were determined using elemental standards in dilute aqueous solution and are based on a 98% confidence level (3 standard deviations).

6.2.4 Selectivity: Sources of Interferences

Interferences in HG may occur either in the liquid phase, during hydride formation and its transfer from solution (e.g., decreased efficiency of hydride formation or changes in the rate of hydride release from the liquid phase), or they can affect the analyte in the gaseous phase, either during hydride transport from the sample solution toward the atomizer or in the atomizer itself.

Interferences in the liquid phase can be due to different reduction efficiencies of different analyte species in the sample or due to the sample matrix. In the first case, adequate precautions should be taken in the sample-preparation stage, like appropriate pre-reduction or sample digestion. For example, pre-reduction is almost indispensable for Se and Te, the higher states of which do not produce hydrides with THB. Also, while in As and Sb determinations both trivalent and pentavalent forms of analytes yield hydrides, pre-reduction is nevertheless highly desirable since arsenite and antimonite are determined with better sensitivity (2– to 8-fold higher). On the other hand, organoelement species of the analyte usually need to be decomposed prior to the HG step, generally by wet chemical attack with strong oxidants at room or elevated temperature (microwave or UV assisted).

Concerning the sample matrix, care should be taken with solid particles capturing analyte ions (e.g., carbon particles from an inefficient mineralization of biological samples), and also with ions of transition and noble metals that form species by reaction with THB capable of capturing the hydride. In most cases, matrix species responsible for such interferences are usually present in great excess compared with the analyte. Then, the extent of the interference should not depend on the analyte concentration but only on the interferent concentration and, therefore, the method of standard additions may be used, in principle, to alleviate these interference effects. In other cases (in particular when the analyte signal is seriously suppressed) interference has to be masked or physically removed prior to the reduction stage.

Other volatile species, most often hydrides but also other compounds, or liquid spray produced in the hydride generator, may produce transport interferences in the gas phase; however, there are not many identified cases of these interferences reported in the literature. Another source of gas-phase interferences is due to interaction and decomposition of the hydride on the surface of the system connecting tubes or even in the atomizer surface itself, very often giving rise to memory effects.

6.3 ELECTROCHEMICAL GENERATION OF VOLATILE HYDRIDES

Although chemical reduction with sodium tetrahydroborate is by far the most common method for HG–AAS determinations, the use of THB for such purposes is not ideal, as some drawbacks of this reagent include (Sturgeon and Mester 2002; Wu et al. 2010): (1) THB is prone to undesired contamination (worsening the limit of detection of analyte determination) and is relatively expensive; (2) its aqueous solutions are unstable and, so, should be prepared prior to use (even if some improved stability may be conferred through addition of 0.1%–2% NaOH, membrane filtration, or refrigeration); (3) the corresponding derivatization reaction may be interfered with by concomitants in the sample solution; and (4) evolving excessive amounts of hydrogen may alter the performance of some detection systems.

Thus, in recent years, electrochemical hydride generation (EcHG) is being studied for analytical purposes (Bolea et al. 2006; Sima et al. 2004). It is important to note, however, that EcHG is not yet fully established for routine analyses by AAS. In a typical EcHG experiment, continuous flows of catholyte and anolyte, respectively, are pumped through cathode and anode channels of an electrolytic flow-through cell. The sample is typically injected into the catholyte stream used as a carrier (e.g., diluted hydrochloric or sulfuric acid). In this way, the hydride is generated in the cathodic space of an electrolytic cell with concurrent oxidation of water in the anodic compartment, while the gaseous reaction products are then separated from the waste aqueous solution in a subsequent gas–liquid separator using a purge gas flow. The main advantage of the EcHG approach is that it does not require any other reagents than diluted acid; this brings also an additional advantage that the risk of contamination is reduced since these acids can be obtained in high purity. Furthermore, the experimental conditions are not as critical, and the influence of the oxidation states of the analyte on the hydride yield can be reduced. It has been also reported that interfering levels in EcHG are lower in general terms than those observed in THB-based generation systems; however, interferences still exist. Using EcHG, the major interference source does not occur in solution, as it does in chemical generation systems, but on the surface of the cathode. As a consequence, masking of interfering ions is usually ineffective in preventing interferences in EcHG systems. Most of the researchers proposing remedial actions for transition metal interference elimination using EcHG have resorted to procedures of physical separation of the matrix and/or pretreatments of the cathode surface (Bolea et al. 2006).

Several cathode materials have been investigated for EcHG. The most useful ones appear to be carbon (glassy or pyrolytic graphite) and Pb (Arbab-Zabar et al. 2006). An often reported construction of electrolytic cell consists of two-half cells, one anodic and the other cathodic, usually separated by an ion exchange membrane or porous glass frit. Figure 6.4 shows a design of a typical electrolytic cell constructed by assembling four blocks, made from Perspex, held together with screws.

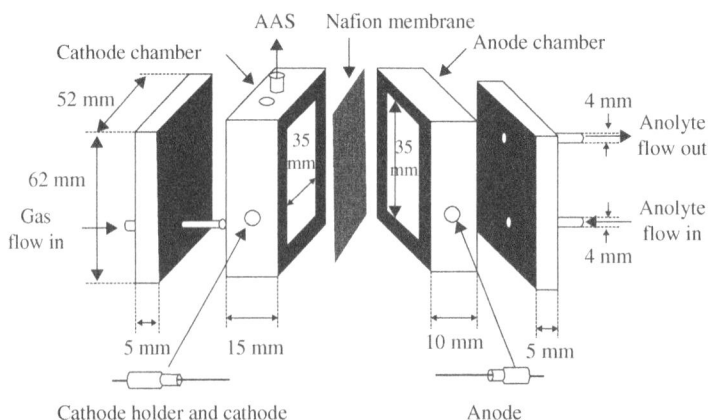

Figure 6.4. Diagram of a electrolytic hydride generation cell for electrochemical generation of volatile hydrides.
Source: Reproduced with permission from Arbab-Zavar (2006) © by Elsevier.

A coil of Pt wire and a lead–tin alloy wire, both having surface areas of 1.0 cm^2, were used as anode and cathode, respectively. An argon carrier gas flow is directly introduced into the cathode chamber for the separation of the generated volatile species. Thin-layer miniaturized flow-through cells with sheet-shaped electrodes have been also proposed (Sima et al. 2004) allowing for low flow rates of catholyte.

EcHG-AAS methods have been successfully applied to trace-level determinations of arsenic, antimony, selenium, tin, bismuth, germanium, and cadmium.

6.4 COLD-VAPOR GENERATION

Since free atomic vapor for most metals cannot exist at room temperature, heat must be applied to the sample in order to break the molecular bonds and to obtain the atoms in the vaporized form needed for AAS measurement. A notable exception to this general fact in nature is mercury: Vaporized free mercury atoms can exist at room temperature and, therefore, mercury can be measured by AAS without the need for a heated absorbance cell. Cold vapor (CV) generation for the determination of mercury was first described in the 1960s, after which it became the leading method for mercury determination. In 1995, two independent research groups (Guo et al. 1995; Sanz Medel et al. 1995) reported on the AAS determination of cadmium following an adequate reaction of the metal aquo-ion in acidic medium with THB to cause CV generation of cadmium for subsequent AAS measurement.

6.4.1 Mercury

Mercury CV is based on the reduction of mercury ions with a reducing agent to generate elemental mercury. Since the vapor pressure of elemental mercury is quite high, Hg^0 can be readily separated from the aqueous matrix with an inert purge gas (e.g., Ar) and then transported and introduced into an absorbance cell. As pointed out above, the analyte reaches the absorbance cell already in atomic form, and so there is no need for an additional atomization step. However, it is usually advisable to heat the cell anyway, to avoid background absorption from water vapor transported by the carrier gas, which at room temperature may condense as small droplets in the absorption cell.

Either tin chloride or sodium tetrahydroborate is used as a mercury reductant. An advantage brought about by the use of tin chloride is that hydrogen gas is not evolved. This is not so important in AAS, but may be of particular concern when plasma-based detectors are used (because hydrogen produces plasma instabilities and so poor precision, for example, in ICP-OES or ICP-MS measurements). However, it should be pointed out that reduction of Hg^{2+} by tin chloride is slower than using THB. This reagent is unable, by itself, to reduce organomercury species; therefore, in this case, the CV determination of total mercury is carried out using a wet digestion process with oxidizing reagents, in order to decompose any organic mercury to Hg^{2+}. Conversely, sodium tetrahydroborate is capable of reducing both inorganic and organo-mercury species. However, the resulting gaseous products are different, with inorganic mercury being reduced to elemental mercury, while the existing organomercury species are transformed to their corresponding volatile hydride species.

6.4.2 Cadmium

Cadmium also has been shown to be able to form monoatomic vapor at room temperature stable enough to be measured by AAS. In fact, CV-AAS has been used advantageously for the determination of Cd in a variety of matrices via reduction of the cadmium ions in solution by $NaBH_4$. It has been reported that the CV reaction efficiency is drastically enhanced in the presence of organic reagents such as thiourea or "organized media," to the extent that reported overall efficiencies for generation of atomic cadmium reached 75%.

The term "organized media" has been applied in Chemistry to describe microscopically organized chemical assemblies, for example, of a surfactant (amphiphile), which spontaneously form within the bulk of a solution. Organized media or "organized assemblies," such as micelles and vesicles, have been shown to exhibit valuable properties in analytical chemistry: They can solubilize, concentrate, and compartmentalize ions and neutral molecules; they can modify equilibria, for example, acid–base and redox constants, reaction rates, chemical pathways; and also influence stereochemistry. Besides, they may drastically change the spectral characteristics of coloured/luminescent compounds (Burguera et al. 2004; Fernández de la Campa et al. 1995).

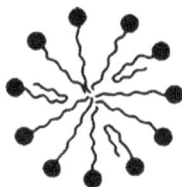

Micelle:
self-aggregation of individual
surfactant molecules
above a critical concentration.

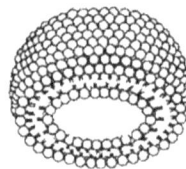

Vesicle:
closed bilayer structure formed by
double-chain surfactants after
adequate sonication.

Although the reaction mechanism is not clear yet, it seems plausible that the reaction intermediate may be CdH_2, which is transportable over significant distances at room temperature, where, under the influence of the UV excitation from the atomic absorption instrument's lamp, it ultimately would decompose to yield the measurable Cd^0 (Sanz Medel et al. 1995).

6.5 TRAPPING/PRECONCENTRATION OF VOLATILIZED ANALYTES

Volatilized metal derivatives can be directly transported to the light path ("direct transfer systems") for analysis. However, the detection sensitivity can be improved by first trapping (and, thus, preconcentrating) the volatilized analyte derivative "in" or "on" an appropriate media or collection device. Then, after VG evolution is finished, the analyte should be rapidly released from the collector. A variety of collection methods were most frequently employed in early years of CV-AAS application when relatively slow chemical approaches for volatile-species formation were employed.

Regarding hydride formation, currently, only cryogenic trapping (located prior the atomization cell) and, most commonly, in situ trapping in a graphite furnace (here the graphite furnace acts also as an atomization cell) are being used. In both cases, hydrogen evolved in the reaction passes freely and, so, is not collected:

1. *Cryogenic trapping*: One of the main advantages of cryogenic trapping compared to other preconcentration methods is that unstable/reactive species do not come into contact with any liquid or solid sorbent material, thereby significantly reducing the possibility of their alteration/decomposition. Typical trapping temperatures are in the –150 to –196°C range, obtained with liquid nitrogen cooling. After the collection stage is completed, analytes are released following fast heat-up.

   However, cryotrapping technology remains rather unpopular: Most cold traps are home made and are operated manually, requiring skill and experience of the operator. In any case, it should be highlighted here that the approach is particularly used online coupled to gas chromatography (GC) in the metal-speciation field, for preconcentration of volatile metal(loid) species obtained by HG/ethylation derivatization, to be subsequently separated by GC.

2. *In situ trapping*: Major impediments to enhancing performance of conventional VG-based analytical techniques are that the species are often diluted by the coevolved hydrogen as well as any carrier gas used to ensure phase separation and transport. Furthermore, some hydrides are not efficiently atomized

in heated quartz-tube sources (e.g., GeH_4). Both these inconveniences can be overcome by resorting to in situ trapping techniques, which couple HG with a graphite furnace and ETAAS determination, thus permitting significant enhancement in detection capabilities over conventional batch and flow-generation procedures. In this approach the graphite furnace (located in the optical path) is used to decompose the volatile hydride and trap the analyte species on the tube surface, or on a surface pretreated (e.g., with palladium or iridium), being later released by a controlled furnace heating program.

In the case of mercury, a collection device based on amalgamation is commonly used to measure samples with very low mercury concentrations (see Figure 6.5): Mercury vapor liberated from the sample aliquot in the reduction step is trapped on a gold or gold alloy gauze (mercury readily forms an amalgam with gold). After a preselected collection time, the gold surface is electrically heated and the absorbed mercury released, the vapor being directed to the absorbance cell (in some designs the mercury amalgamation trap is fitted into the quartz tube). In principle, the only limit to this approach would be that imposed by background or contamination levels of mercury in reagents or system hardware. Presently, several companies manufacture compact, automated atomic absorption instruments dedicated to the determination of mercury, with the amalgamation collection system as an option.

6.6 APPLICATIONS AND EXAMPLE CASE STUDIES

Both hydride- and CV-generation accessories are today commercially available and relatively inexpensive, offering low DLs, possibility of automation (in

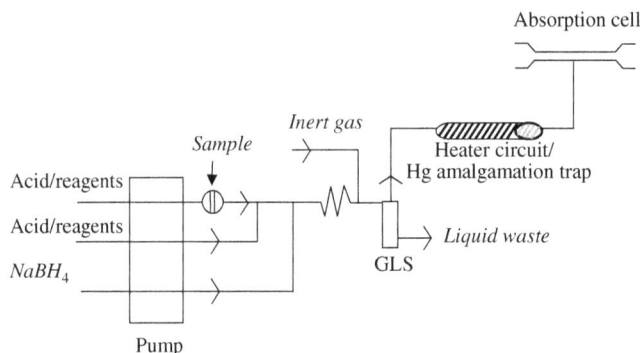

Figure 6.5. Schematic diagram showing the key components of a flow-injection system with mercury amalgamation trap assembly for mercury determination by CV.

flow modes), and good sampling frequencies (e.g., 40–60/h). Furthermore, as shown in Table 6.2, HG offers certain (though limited) potential for analytical metal speciation.

On the other hand, HG and CV methods are not free from intrinsic limitations and drawbacks. A major limitation is that, in practice, they are restricted to only a few analytes. Also, experimental parameters should be carefully controlled and care needs to be taken with chemical interferences. Specific chemical pretreatment is often needed so as to reliably transform the analyte element into a definite, reactive chemical form and oxidation state (Tsalev 2000). For example, selenium and tellurium analytes have to be pre-reduced to their hydride-forming oxidation states, Se(IV) and Te(IV); lead should be pre-oxidized to its metastable Pb(IV) form; arsenic and antimony usually exhibit much better sensitivity as their inorganic As(III) and Sb(III) oxidation states; therefore a pre-reduction step for samples containing higher As and Sb oxidation states is often necessary.

Today, HG-AAS for As (Anawar 2012) and CV-AAS for Hg (Niece et al. 2013) are both well-established, highly competitive techniques for these two otherwise difficult analytes. To illustrate these two typical AAS determinations, where improved sensitivity is achieved by resorting to volatile-species generation, the determination of arsenic in waters and of mercury in hair will be described below in more detail. On the other hand, in situ trapping of a hydride in a preheated graphite tube is an interesting strategy to achieve high analytical sensitivity, by combining the advantages of HG-AAS and ETAAS. More recently the coupled HG-ETAAS has become a rather common analytical tool. Therefore, the determination of total selenium in beans and soil by HG-ETAAS combination is finally described (Shaltout et al. 2011).

Table 6.2. Examples of HG-AAS speciation potentialities.

Species determined	Differentiation from	Principle
i-Se(IV)	i-Se(VI)	Reaction kinetics
i-Te(IV)	i-Te(VI)	Reaction kinetics
i-As(III)	i-As(V), MMA, DMA	pH; $NaBH_4$ concentration
i-As(III) + DMA	i-As(V), MMA	pH
i-As(III) + i-As(V) + MMA + DMA	AsBet, Me_4As^+, arsenosugars	Reaction kinetics

DMA: dimethylarsinic acid; HG-AAS, hydride-generation atomic absorption spectrometry; i: inorganic; MMA: monomethylarsonic acid.
Source: Adapted with permission from Tsalev (2000). Reproduced with permission from Elsevier.

6.6.1 Determination of Arsenic in Waters

Human exposure to arsenic can cause a variety of adverse health effects. Living organisms in general may be exposed to toxic arsenic species primarily from food and water. The contamination of groundwater with arsenic has been reported in several countries around the world. In particular, 50 districts of Bangladesh and nine in West Bengal (India) have arsenic levels in drinking water well above the guideline values ($10 \ \mu g/L$) provided by the World Health Organization (WHO). According to WHO, in the year 2000, "the contamination of groundwater by arsenic in Bangladesh is the largest poisoning of a population in history." Hyperkeratosis on the palms and feet is the common symptom of arsenic poisoning. Long-term exposure to low concentrations of arsenic has been reported to cause cancer of bladder, skin, and other internal organs.

In groundwater, arsenic is predominantly present as arsenite and arsenate with a minor amount of monomethylarsonic acid and dimethylarsinic acid (see corresponding formulae in Figure 6.6). Under appropriate experimental conditions all four species can be reduced to As volatile hydride with THB, although transition metals might interfere with this sensitive determination of arsenic. The predominant mechanism is probably due to the reaction of the interfering transition metal ion with $NaBH_4$ reductant, forming a precipitate that is able to encapsulate and/or catalytically decompose the hydrides. Generally, L-cysteine has proved to be very useful for preventing iron interferences, commonly present at high concentration in many types of samples. It was found that a cysteine–THB intermediate formed showed higher reducing power than THB alone.

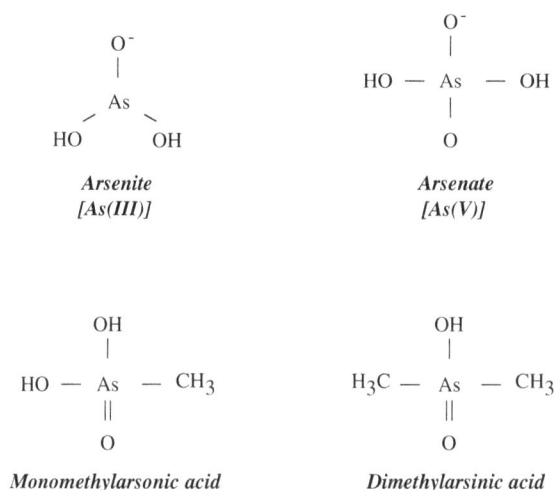

Figure 6.6. Arsenic species in water.

HG can be also used for differential determination of As(III) and As(V), just by taking advantage of the fact that As(III) reacts with THB at a higher pH than As(V). It should be noted, however, that the analytical potential of this approach for chemical speciation in general, and for As speciation in particular, is rather limited. When speciation of all possible arsenic compounds present is required, the combination of a more powerful separation method (e.g., chromatography) prior to HG-AAS is mandatory (see Chapter 7). Speciation information on arsenic in natural waters is very important today because the bioavailability and physiological and toxicological effects of arsenic will depend on its chemical form (for instance, arsenite is 20–60 times more toxic than arsenate and several hundred times more toxic than methylated arsenic compounds).

The lack of long term stability of arsenic species in water samples is a crucial aspect in As and other speciation studies. Procedures used for sample storage, handling and analysis may result in final oxidations of arsenite to arsenate. In ground water, As(V) has commonly been reported as the predominant water-soluble As species. However, recent studies have shown that As(III) also may be present in high concentrations in certain waters. Therefore, particular attention should be paid to potential changes of sample conditions when transporting it form the field to the laboratory and during storage, because alteration of chemical species in the original sample might happen changing the original species present in the water. If no reliable sampling and storage conditions can be found, on-site analysis with portable instruments may be necessary.

6.6.2 Determination of Mercury and Methylmercury in Hair

The first well-documented cases of severe poisoning by mercury compounds are available from Minamata Bay (Japan) in 1956 when sea fish were severely contaminated with methylmercury (coming from an industrial spill). Then, other episodes of poisoning with this compound were demonstrated around the world, for instance, when mercury-contaminated seeds (wheat treated with a methylmercury fungicide) were used in Iraq in 1971 to prepare bread. Subsequent studies to identify the nature of such episodes demonstrated that direct intake of water or food contaminated by methylmercury has extreme detrimental effects on the human central nervous system. Over the past three decades, contamination of the food chain by organomercury species has become a worldwide concern with increased public awareness. For example, studies in the Brazilian Amazon demonstrated that the Indian population had increased exposure to methylmercury because of their consumption of fish contaminated by upstream gold-mining activities. Elemental mercury used in such activities undergoes natural methylation reactions in the water stream to be converted to soluble inorganic and methylmercury species. This methylmercury is able to bioaccumulate in small fish and then subsequently occur all along the food chain to reach the human population. Methylmercury is considered to be 50–100 times more toxic than Hg^{2+}. Thus, determination of total mercury and methylmercury (i.e., analytical speciation of the metal) is needed today.

Hair is a suitable indicator for monitoring of human exposure to mercury and methylmercury because hair reflects both recent and historical exposure to mercury (time-based exposure of 1–2 years with long-haired individuals can be monitored). Also, the use of hair as a mercury-control sample for possible poisoning offers practical advantages (a hair sample is easier to collect and to store compared with blood or organ tissues). As a general rule, detection and final determination of mercury are carried out to the required sensitivity by CV-AAS. A previous sample pretreatment step is recommended in order to clean the hair samples (e.g., acetone or a nonionic surfactant such as Triton X-100), followed by a final washing with redistilled water and then drying at 105°C.

For total mercury determinations, weighed clean hair samples are heated in a concentrated acid medium for sample mineralization and made to volume in a volumetric flask before the final CV-AAS measurement.

While the high volatility of atomic mercury is beneficial for its highly sensitive determination, keep in mind that analyte losses may result during sample mineralization. This is why enclosed digestion bombs, heated by either convection or microwave irradiation, are usually employed for total mercury digestion. Alternatively, wet-digestion on a steam-bath gives satisfactory results as well, provided that oxidation chemistry is adjusted so that the organic matrix is oxidized, and all the released mercury is converted into nonvolatile Hg^{2+}.

For methylmercury species determinations in hair preserving the methylmercury compound is critical; it must be considered that it is usually bound either to cystine sulfur or to sulfhydryl groups present in protein amino acids. Therefore, for the determination of the highly toxic methylmercury it is mandatory that a special digestion step is undertaken, which is able to break the existing bonds between hair and the analyte while preserving intact the chemical nature of methylmercury.

6.6.3 Determination of Selenium in Bean and Soil Samples Using Hydride Generation—Electrothermal Atomic Absorption Spectrometry

Selenium is an essential micronutrient for humans and animals. It has an important role as a substance with anticancer properties and probably for heart disease prevention. The U.S. Food and Nutrition Board established dietary reference intakes for selenium (e.g., 15 and 20 µg/day for infants and 70 µg/day for lactating mothers). An upper limit for safe intake was set at 400 µg/day.

Various plants growing in selenium-rich soils absorb and accumulate selenium. Because of the importance of this element, geographical selenium maps are already available for different regions in the world. Selenium uptake depends on its chemical form (speciation). Selenomethionine has been shown to be the predominant form of selenium in proteins from wheat, soybeans, and selenium-enriched yeast.

ETAAS and HG-AAS are frequently used for this analytical purpose. While the response of HG-AAS is strongly dependent on the selenium species present, ETAAS is adequate for the determination of total selenium. In HG-AAS the incomplete mineralization of refractory organic selenium compounds such as selenomethionine is one of the major challenges, whereas selenium losses by volatilization were often the cause for erroneous results in ETAAS. Furthermore, there are nearby iron atomic emission lines that might cause spectral interference of the most sensitive selenium line at 196.03 nm when conventional line-source AAS is used. This line is also in the range of strong molecular absorption due to PO and NO bands, which might cause spectral interferences (derived from under- or overcorrection in ETAAS). The combination of HG with ETAAS can therefore be a good choice to overcome some of the problems of Se determinations by ETAAS while still achieving very good sensitivity.

The first step of this HG-ETAAS methodology is the generation of the gaseous SeH_2 and this requires selenium to be present as inorganic Se(IV), that is, as selenite. This means that any organic selenium compounds have to be mineralized first. This might require quite harsh conditions. All Se(VI) has to be reduced to Se(IV) prior to HG. Comparison of several digestion methods for Se determination in biological samples has shown that only the HNO_3 + $HClO_4$ or HNO_3 + H_2SO_4 + H_2O_2 mixtures yielded complete recovery of Se. Microwave-assisted wet digestion based on a mixture of HNO_3 and H_2O_2 followed by UV irradiation can be a successful approach as well, provided that after digestion the solution is boiled with 5 or 6 mol/L HCl under reflux for 15–30 min to reduce all hexavalent selenium to tetravalent selenium.

For ETAAS, coating of the pyrolitically coated graphite tube with iridium as a permanent modifier for the trapping of selenium (and other hydride-forming elements) in the atomizer was found to be a good choice as such tubes could be used for several hundred measurements. The use of iridium coating also gives rise to a high efficiency of the selenium hydride deposition.

The graphite furnace temperature program for trapping the H_2Se consisted of a 30 s collection stage at 500°C with an argon flow rate of 300 mL/min.

This step was followed by a 5 s atomization stage at 2,100°C without Ar flow. In order to avoid loss of the permanent modifier at high temperatures, the cleaning temperature (3s with 300 mL/min of Ar) was set at 2,200°C, which was sufficient as no significant residues were expected in the graphite tube using the HG methodology (Shaltout et al. 2011).

7

Flow Analysis and Atomic Absorption Spectrometry

Flow-based methods for sample manipulation offer several important practical advantages and analytical performance improvements as compared to manual-batch alternatives. In this chapter, after an introduction to the basic philosophy of the flow-injection analysis—atomic absorption spectrometry (FIA–AAS) "hyphenated" approach, the main components of the flow devices will be described. Simple tools of the trade, such as the injection valve and the peristaltic pump, as well as strategies for two-phase sample preconcentration and devices for convenient sample digestion are also presented and their use illustrated. In addition, the online coupling of chromatographic separations (which can be considered also as flow systems) to atomic absorption spectrometers is then described in this chapter.

Finally, the analytical potential of such combined or coupled analytical techniques is highlighted by illustrating their use in solving three selected analytical problems by resorting to flow analysis—AAS.

7.1 INTRODUCTION

Flow-based methods based on microfluidic manipulation of samples and reagents are widely used to carry out a wide variety of sample pretreatments online coupled with a detector system. The basis of such methods relies on three aspects that should be kept reproducible: (1) volume of sample injected into the flow manifold; (2) dispersion of the injected sample throughout the conduits of the manifold (precise pumping is therefore required); and (3) timing for each sequence carried out in the flow system (i.e., mixing with other streams, time to reach the detector, etc.). In this way, every sample and standard is subjected to an identical treatment and is measured after the same time interval, so there is no need to wait for complete reactions.

Sample manipulation flow-based methods bring about several important advantages when compared to manual-batch counterparts, including: lower risk of sample contamination, possibility of using unstable reagents and no need to obtain stable products, low sample volume required, high sample throughput, and possibilities for online preconcentration and/or elimination of interferences. The use of flow procedures allows for the improvement of analytical properties like precision, possibility of online sensitivity enhancement and elimination of interferences, reliable indirect analysis, microsample analyses, and low-cost analysis (in terms of personnel, time, and reagents).

Automated sample flow processing can be based either on continuous flow or on programmable flow. Continuous-flow techniques, such as flow-injection analysis (FIA), use constant forward motion of the carrier to transport the sample from injector to detector. The name of FIA was coined by Ruzicka in 1975 and the first reports on methodologies explicitly defined as FIA for atomic absorption spectrometry (AAS) were published in 1979 (Wolf et al. 1979; Yoza et al. 1979). In FIA techniques, samples are injected into a carrier solution, which then transports the sample zone toward a detector during which desired (bio)chemical reactions take place (typically, the analyte carrier stream and a stream of reagent are mixed at a confluence point). The mixed stream flows constantly through the detector giving rise to a transient signal. As can be seen in the scheme of Figure 7.1a, FIA allows for high visual observation of the needed analytical processes before final detection by AAS.

In FIA analysis, the profile "concentration versus time" (FIA curve or fiagram) observed at the detector, for an injected sample with an analyte concentration Co, is a peak. This peak is the result of the dispersion processes occurring to in the sample during the flow. The ratio of the injected analyte concentration, Co, to the instantaneous concentration corresponding to any point of the sample volume reaching the detector, Cg, is known as the dispersion coefficient, Dg. The value of the dispersion coefficient at the peak maximum, D, can be used to characterize the extent of mixing in a given FIA manifold. Thus, in most FIA experiments, D values are higher than unity due to dispersion, but if a concentration step has taken place, the resulting D values are lower than unity.

Programmable flow techniques, such as sequential-injection analysis (SIA), use flow reversals and flow acceleration to mix the sample with reagents. SIA techniques were first developed in 1990 (Ruzicka 1990). In a SIA manifold, sample and reagent solutions are sequentially aspirated into a holding coil and

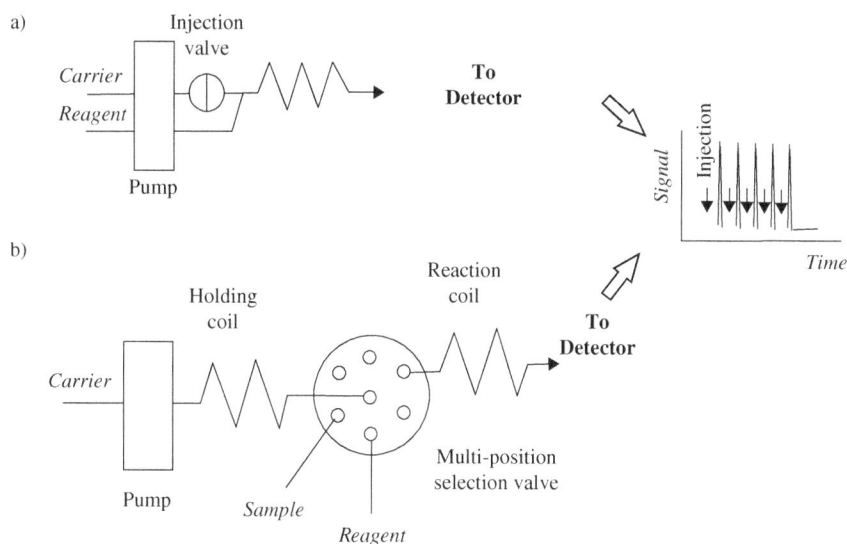

Figure 7.1. Schematics of flow injection and sequential injection manifolds. (a) Basic two-channel flow-injection analysis (FIA) manifold. (b) Simple sequential-injection analysis (SIA) manifold.

stacked, the aspirated volumes being determined by the time and aspiration rate; the mixture is promoted by flow reversal, while sending the stacked zones toward the detection system. The computer control ensures the reproducibility of the process. Sequential injection has the ability of performing different determinations without system reconfiguration (placing different reagents on the ports of the selection valve) and there can be a reagent saving associated with noncontinuous consumption.

The heart of an SIA manifold is a multiposition-selection valve (Figure 7.1b). Fluids are manipulated within the manifold by means of a bidirectional pump, and accurate handling of sample and reagent zones necessitates computer control. A holding coil is placed between the pump and the common port of the multiposition-selection valve. The selection ports of the valve are coupled to sample and reagent reservoirs as well as to a detector (the placement of different reagents on the ports permits different determinations with the same manifold). In a first step the valve is directed to the selection port that is connected to the sample line and a zone of the sample is drawn into the holding coil by the pump. Then, the selection valve is directed to a port that is connected to a reagent line and a zone of the reagent is drawn up into the holding coil adjacent to the sample zone. Finally, the selection valve is switched to a port which is connected to a detector. As the zones move through the reaction coil toward the detector, zone dispersion and overlaps occurs.

The majority of SIA procedures are based on solution-phase chemistry, although in recent years (Economou 2005) the scope of SIA has been extended to more complex, online sample-manipulation processes. However, considering the broader application fields and the easier "visualization" of FIA procedures, attention will be focused to FIA-based methods in this chapter.

On the other hand, the online coupling of chromatographic separations (flow systems) to AAS is also rather straightforward and most useful. The information given by such a powerful combination is unique, in particular to solve element speciation problems. For this reason, Section 7.9 discusses the coupling of chromatographic separation techniques to atomic absorption spectrometers.

7.2 FLOW INJECTION ANALYSIS AND ATOMIC ABSORPTION SPECTROMETRY

As illustrated in Figure 7.2, there is a great variety of sample manipulation processes that can be carried out by flow operation procedures coupled to atomic

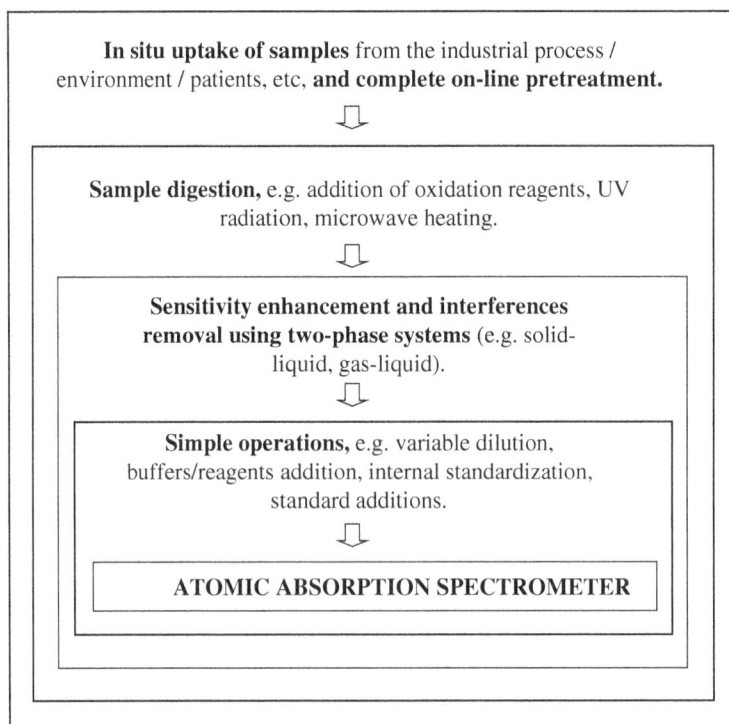

Figure 7.2. Scope of the sample pretreatment processes covered by online flow strategies.

detection. These processes can be classified according to a hierarchy going from simple operations, such as online dilution and standard additions, to pre-concentration/separation stages using two-phase systems, through to sample digestion and even in situ uptake of samples. Although a higher degree of manifold complexity is expected the larger the applications square depicted in Figure 7.2, it is clear that flow injection systems offer high versatility, allowing for the opening of new automation paths to facilitate routine analysis by AAS.

Processes like dilution, reagent mixing, or standard additions can be considered as the more straightforward pretreatment operations, and robust online flow manifolds to tackle these sample pretreatment stages are now routinely used. Flow systems can easily integrate non-chromatographic separation/preconcentration techniques online with the detector. Such separation/preconcentration techniques are based on the use of two phases (solid–liquid, gas–liquid, or liquid–liquid) to obtain two sample fractions, one of them containing the enriched analyte(s) free from potential matrix interferences and the other containing the matrix.

Online flow systems for the decomposition/dissolution of solid samples constitute a further extension of sample pretreatment procedures enhanced by the use of flow manifolds. In this direction of development, different approaches including online chemical oxidation, photo-oxidation, and microwave heating are now successfully exploited. Finally, the in situ uptake of samples directly from their source (industrial process, environment, or even medical patients) and their complete online pretreatment before atomic detection can be considered as the ideal to be pursued in terms of automation. This goal has been addressed and interesting manifolds have been proposed to wholly automate AAS analytical measurements in clinical laboratories (e.g., the procedure starts with the in vivo sample uptake of blood samples), for in situ environmental analysis, as well as for industrial process control.

7.3 BASIC INSTRUMENT COMPONENTS: SAMPLE INTRODUCTION UNIT, PROPULSION SYSTEM, AND CONNECTING TUBES

The basic elemental units of common flow manifolds are described in this section. Examples of integrated flow systems for simple analytical operations are discussed in Section 7.4.

The three basic components of a flow system are: the sample-injection port, the propulsion unit, and the connecting tubes. In addition, special components allowing for on-line separation/preconcentration are also frequently introduced into the manifold and they will be detailed in Sections 7.5 to 7.7, while the setups used for sample digestion are dealt with in Section 7.8.

7.3.1 Sample Introduction Unit

An injection port inserted in the flow system is used for sample introduction in the FIA manifold. The injection port should allow for precisely introducing a volume of sample as a plug into the continuously moving carrier stream in such a way that the movement of the stream is not disturbed.

In the early stages of FIA development, the liquid samples were manually introduced with a syringe through a rubber septum. However, this was far from ideal since among the reported problems it was observed that a lack of reproducibility in the manual injection and components' leaching from the septum material after repeated injections on the same spot occurred. Since then, different sample-introduction approaches have been investigated, for example, valveless injection systems in which the sample introduction is time controlled or volume controlled. Figure 7.3 shows schematically a commonly used six-way sliding rotary valve. This commercially available volume-based injection port has an external loop. In the "load" position the loop is filled with the sample while in the "injection" position the content of the loop is transported through the manifold by the carrier solution. This external loop has to be replaced if a different sample volume needs to be analyzed. A decrease of sample volume or an increase of the mixing coil dimensions will give rise to higher dispersion of the sample bolus.

7.3.2 Propulsion System

The propulsion unit should provide a continuous and reproducible flow rate of carrier solutions through the manifold. Considering that in some cases more than one solution has to be propelled, multichannel capabilities are advisable. Besides, the ideal delivery system should allow for selecting the appropriate flow rate and provide a pulse-free flow. Three main types of propulsion

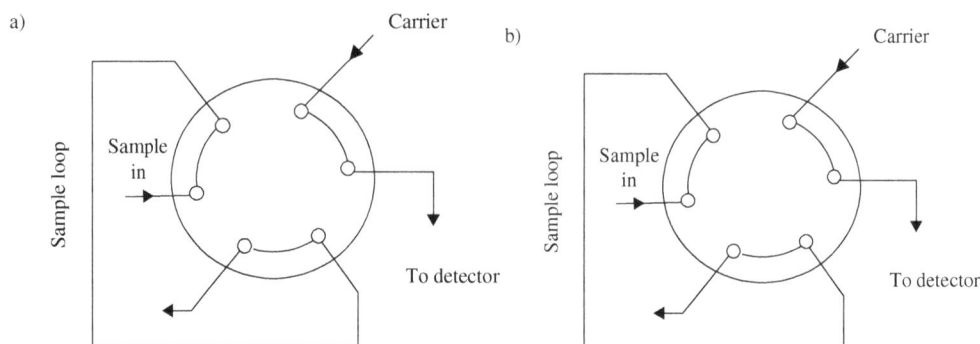

Figure 7.3. Diagram of a six-way rotary valve. (a) Load position. (b) Inject position.

strategies have been proposed for the combination of flowing systems with atomic detectors:

1. A propulsion-less manifold, which relies on the negative pressure generated by the nebulizer to draw the liquid solution to the instrument.
2. The use of a gas-pressurized carrier reservoir to propel solutions through the manifold. The gases commonly used are air or nitrogen. Although this system gives rise to a pulse-free flow, drawbacks such as the consumption of gases or the difficulties to achieve constant flow rates have greatly limited its use.
3. Electrical pumps (such as peristaltic pumps and syringe pumps), which are by far the most common propulsion systems.

Figure 7.4 shows the schematics of the pumping principle of a peristaltic pump. The flow rate depends on the rotation speed and the inner diameter of the pumping tubes. A wide variety of flexible tubes with different inner diameters and made of different materials are commercially available (e.g., Tygon®, poly(vinyl chloride) (PVC) or PVC derivatives, silicone rubber), which can be selected depending upon the solution to be pumped (e.g., acids, organic solvents, etc.). Peristaltic pumps tend to give rise to small pulses that can be significantly reduced by an adequate adjustment of the clamps pressing the pumping tubes.

The syringe pump is a large, electrically operated simulation of a hypodermic syringe. Such syringe pumps can be used even if the backpressure is relatively moderate (e.g., backpressure due to the incorporation of a capillary or a packed minicolumn in the flow system). Syringe pumps exhibit higher precision than peristaltic pumps when a microflow delivery is required.

7.3.3 Connecting Tubes

Inert and flexible tubing with inner diameters in the interval 0.3–1.5 mm allows for the connection of the different units of the flow manifold. In some

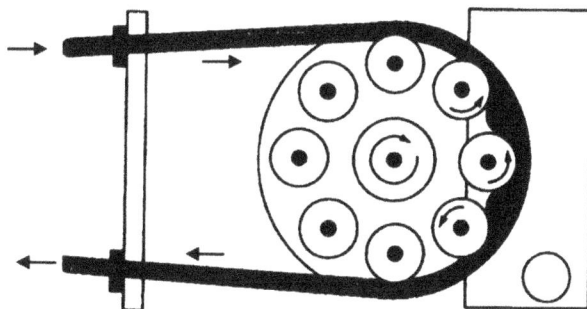

Figure 7.4. Head of a peristaltic pump.

parts of the system the tubing is bent with a coiled shape to allow for an efficient mixture of reagents and sample. Multiway connectors are frequently used to mix flows from different channels.

7.4 SIMPLE COMMON MANIFOLDS: DILUTION, REAGENT ADDITION, AND CALIBRATION

Many types of manifolds can be mounted using just the components described in the previous section, and some are commercially available as accessories for atomic absorption spectrometers. To illustrate such versatility, Figure 7.5 shows five diagrams of flow manifolds. The system shown in Figure 7.5a is the simplest one (single-line manifold). Apart from the obvious advantages of using this system (e.g., analysis of small sample volumes, high sample throughput), its combination with AAS allows the straightforward analysis of samples with high viscosity or with important dissolved solid content, since blockage of the nebulizer or the burner is minimized. On the other hand, this single-line manifold has been proposed also for standard addition using a configuration called "reagent-injection FIA"; in such configuration, the sample is used as the carrier stream into which standards are injected.

The addition of a second channel (Figure 7.5b) allows for carrying out some sample manipulations or pretreatments, such as enhanced dilution, addition of reagents and masking agents (e.g., introduction of caesium or lanthanum in F AAS). The flow system depicted in Figure 7.5c (called "merging-zones"), which contains an injection valve for the sample and a parallel injection valve for reagents, enables important savings in reagent consumption as compared to Figure 7.5b configuration.

As was pointed out earlier, the passage of the sample-dispersed zone through the detector produces an FIA transient signal proportional to the analyte concentration in the sample, usually recorded as a peak. For most applications, peak heights (associated with the most concentrated portion of the dispersed zone) or peak areas are measured. However, measurements related to other portions of the dispersed zone, where the concentration gradients are more pronounced, offer also information of analytical interest. Exploitation of gradients has expanded the range of applications of FIA, allowing calibrations based on a single standard solution and selectivity evaluations, inter alia. However, the precision of measurements may deteriorate with fluctuations in flow parameters, the effect becoming more severe when the measurements are carried out in a region of the dispersed zone with pronounced concentration gradients. Manifolds presented in Figures 7.5d and 7.5e have been proposed to overcome such effects.

The setup shown in Figure 7.5d allows for achieving a variety of dilution factors with just a single injection. The manifold consists of using tubing with

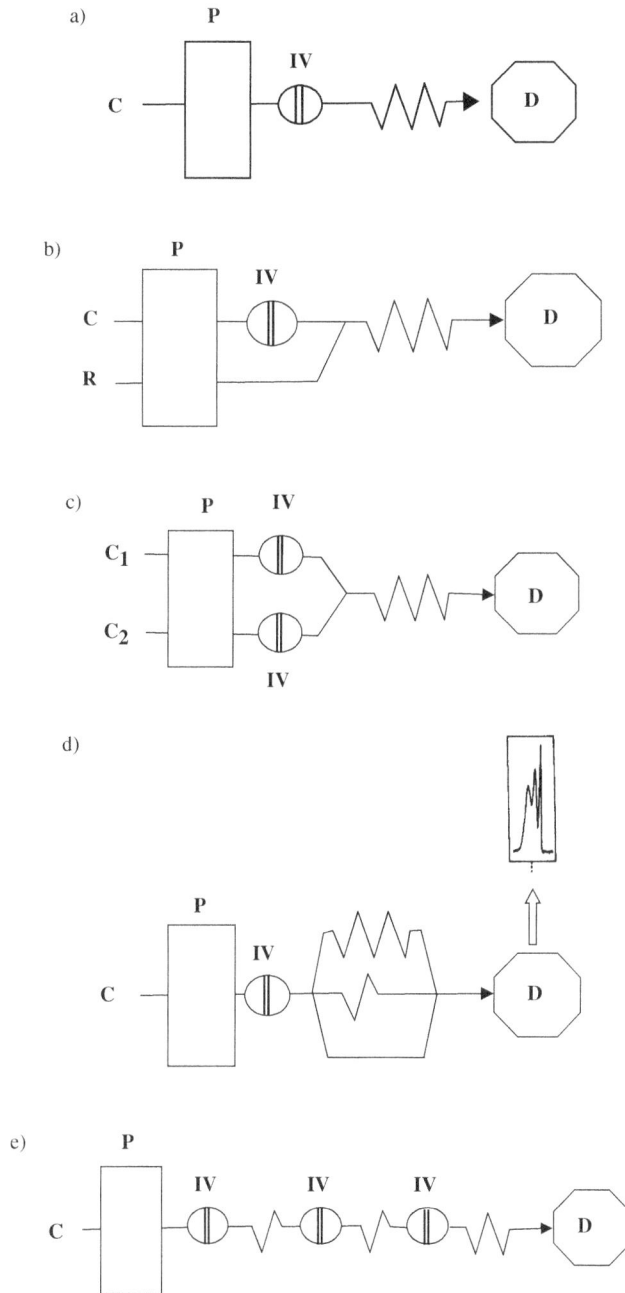

Figure 7.5. Selected examples of simple flow operation systems. P, peristaltic pump; IV, injection valve; C, carrier; R, reagent; D, detector. (a) Single line; (b) multi line; (c) merging zones; (d) three-branch online sample dilution system; (e) simultaneous sample injections flow system.

different lengths, so that the residence time of the subsampled zones in each branch is different. Multiple peaks are formed on recombination. In general, for *n* unequal branches, *n* peaks are produced. For example, Figure 7.5d illustrates a three-branch manifold that could give rise to three partially overlapped peaks (with five possible measurement points, three maxima and two minima). The method generates a family of calibration graphs of varying sensitivity.

In the manifold of Figure 7.5e the required measurements are made in portions of the dispersed zone in which the concentration gradients are minimized. To operate the system, several plugs of a given sample solution are simultaneously injected through injection valves into the same carrier stream, allowing them to overlap as a consequence of dispersion. Provided that plugs of the same solution are used, their merging results in a final dispersed zone characterized by several sites without appreciable concentration gradients, corresponding to the maximum and minimum values of the concentration/time profile. The measurements related to those sites offer a high reproducibility because they are less affected by fluctuations of the flow parameters.

7.5 SOLID–LIQUID SEPARATION AND PRECONCENTRATION

Solid–liquid separation manifolds may be classified according to the separation principles and media used for retention: (1) sorption on a solid phase packed inside a minicolumn (typically uniformly bored or with a conical shape), which is inserted into the flow system; (2) precipitation and coprecipitation, based on the combination in a flow of precipitation of the analyte, its filtration, and final dissolution (in comparison with precipitation methods, coprecipitation is less demanding in terms of solubility of the formed precipitate).

7.5.1 Sorption

Many sorbents have been developed and some of them are marketed as prepacked minicolumns. The characteristic features of online preconcentration flow systems demand some specific properties of the packed material that may be only of minor importance in batch or traditional column procedures: In flowing systems the solid phase has to be reusable and the kinetics of processes or reactions have to be fast (analytes have to be efficiently retained and readily eluted). Besides, in order to avoid backpressure and nonuniform flow patterns it is important that the packing material does not swell or shrink when passing the required solutions (e.g., carrier and eluent).

A plethora of solid phases have been described in the literature for the temporary retention of low-abundance metal ions and charged metal chelates.

Sorption preconcentration strategies reported can be divided into two general groups (see Figure 7.6).

1. Analyte ions are collected directly by a selective solid phase and then eluted from the solid with an appropriate eluent. The sorption of ions involves the use of solid phases (containing a suitable ion exchanger, a chemical reagent, or even microorganisms) immobilized on the solid support packed in a minicolumn. The solid active phases can be purchased (e.g., Chelex 100, which consists of iminodiacetate groups on a styrene divinylbenzene support, or 8-hydroxyquinoline bound to controlled pore glass) or prepared at the working laboratory. An extensive review by Camel (2003) containing more than 250 references about solid-phase extraction of trace elements has been published. Eluents commonly used are acids, bases, or complexing agents. To show the practical utility of this approach, Section 7.10.1 illustrates the sensitive determination of aluminium in dialysis concentrates by using an online minicolumn.

2. Metal ions are adsorbed on suitable solids as previously formed hydrophobic metal chelates (formed online in the flow system) to be later eluted with a hydro-organic solvent. Sorbent materials include, for example, activated carbon, octadecyl functional group–bonded silica gel, modified copolymers, and nanomaterials having properties such as a large surface area to volume ratio, high chemical and thermal stability, and excellent mechanical strength (metal-oxide nanomaterials and carbon derivatives, such as graphene or carbon nanotubes, are among the nanomaterials proposed) (Miró and Hansen 2013). The most commonly used chelating reagents are dithiocarbamate derivatives, which possess very active sulfur metal-binding sites.

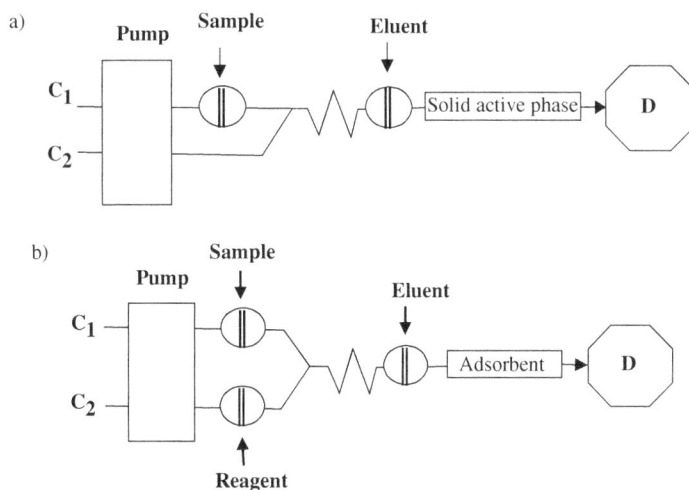

Figure 7.6. Basic manifolds for online preconcentration of a solid phase. (a) System for sorption of ions. (b) System for flow metal–chelate formation and subsequent sorption.

This latter approach is of particular interest when reagents with slow binding kinetics to the metal analyte are used. It is appropriate when the bond between the metal and the reagent is too strong, since there is no need to break down these bonds during elution. However, it has some disadvantages: the sorbent support can become saturated with the free reagent (therefore the efficiency of metal chelate retention is decreased) and the hydro-organic character of the solvents used as eluents can cause problems with some atomic detectors.

The compatibility of organic solvents with electrothermal atomic absorption spectrometry (ETAAS) has prompted a number of interesting applications based on the sorption of metal chelates on nonionic sorbents and their later elution with organic solvents such as ethanol, methanol, and acetonitrile.

7.5.2 Precipitation and Coprecipitation

Continuous precipitation–filtration–dissolution processes can be considered as solid–liquid flow systems in which the second phase is first generated and then disappears in situ. Precipitation and coprecipitation flow systems have some resemblance to metal–chelate sorption (a filter replacing the sorbent column), except that no column capacity limitations exist for precipitation/coprecipitation methodologies.

Coprecipitation systems have been synchronously coupled to ETAAS for the determination of trace amounts of heavy metals. The analytes are coprecipitated with voluminous agents, such as the iron(II)-hexamethylenedithiocarbamate complex, on the walls of a knotted reactor. The precipitate is later dissolved in a few microliters of isobutyl methyl ketone, stored in a poly(tetrafluoroethylene) (PTFE) tube and delivered finally into the graphite-tube atomizer for AAS measurement.

7.6   GAS-PHASE FORMATION STRATEGIES

As already explained in Chapter 6, mass transfer between a liquid initially containing the analyte and a gas phase, which becomes enriched by the analyte (or a gaseous derivative), offers some interesting features in combination with atomic detectors. In such systems, 100% of the analyte in gaseous form is introduced into the atomic detector (as compared to 1%–5% of the analyte introduced by conventional liquid nebulzation) giving rise to important sensitivity improvements. In addition, a striking suppression of interferences during atomization is also achieved.

7.6.1 Flow Systems for the Formation of Volatile Derivatives of the Analyte(s)

The most popular volatile compounds used in combination with AAS are covalent binary hydrides and mercury cold vapor. Despite its well-known advantages, classical batch mercury cold-vapor or hydride-generation procedures have a number of pitfalls, which can be overcome with the use of flow systems, including: (1) relatively modest throughputs; (2) poor precisions due to incomplete control on reaction conditions, particularly reaction time; (3) the common hydride-generation process used also generates hydrogen in excess, which, if released as a sudden burst, could affect the flame; and (4) quite large amounts of sample are needed. Such drawbacks can be reduced by the use of flow procedures. However, on the other hand, such use of flow systems will require rapid analyte volatilization kinetics (e.g., fast chemical reactions) and efficient mass transfer between both phases (the co-occurrence of the two phases in a hydrodynamic system gives rise to some problems that call for ingenious technical solutions).

The earliest continuous hydride-generation system connected online to AAS was reported by Aastroem (1982). Since then, online methods for mercury cold vapor and covalent hydride generation have become well established. Among other analytically useful volatile compounds, the significant analytical potential demonstrated by the generation of cadmium cold vapor (Sanz-Medel et al. 1995) and alkyl derivatives of several metal(loids) should be highlighted here (Rapsomanikis 1994).

In continuous hydride- or cold-vapor-generation systems the sample is mixed online with the reducing agent. The volatile species formed are separated from the liquid in a gas–liquid separator (GLS) with the aid of a continuous flow of a stripping gas. The stripping gas and the volatilized species (analyte compounds + by-products continuously produced such as H_2, CO_2, or water vapor) are continuously swept to the detector, while the liquid is driven to waste. Argon, at a constant flow rate generally between 50– and 500 mL/min, is most frequently used as the stripping gas.

Besides the factors affecting the sensitivity of batch-hydride- or cold-vapor-generation methods (e.g., matrix components, concentration of reagents, and type of detector) the following operational parameters have to be also considered for continuous systems: the flow rate of the sample channel, the ratio of flow rate sample/reducing agent, the flow rate of the stripping gas, and the design of the GLS. Separators to be used in continuous systems should meet two principal requirements: (1) they should work smoothly and regularly in order to avoid irreproducibility and to decrease signal fluctuations, and (2) they should induce minimal dispersion or dilution of the analyte in gaseous form, that is, they should have dead volumes as low as possible, though not so low as to give rise to incomplete gas–liquid separation.

Many different designs have been evaluated as GLSs, some of which are shown in Figure 7.7 (each is not drawn at the same scale). The most common designs consist of chambers made of glass with proper entrance and exit orifices. The use of gas-permeable membranes (in connection with flat or concentric hollow cylinders) has also been proposed to separate gases from liquids in flow systems; for example, some practitioners have utilized a microporous PTFE membrane as a diffusion medium for the separation of the gaseous analyte derivative from solution. Figure 7.8 depicts a typical FIA manifold to achieve online gas–liquid separation.

Most manufacturers of atomic absorption spectrometers are also marketing flow-injection systems for cold vapor/hydride generation, as an AAS accessory. Furthermore, compact atomic absorption spectrometers specifically dedicated to the fully automated and fast determination of mercury using a flow system are also available.

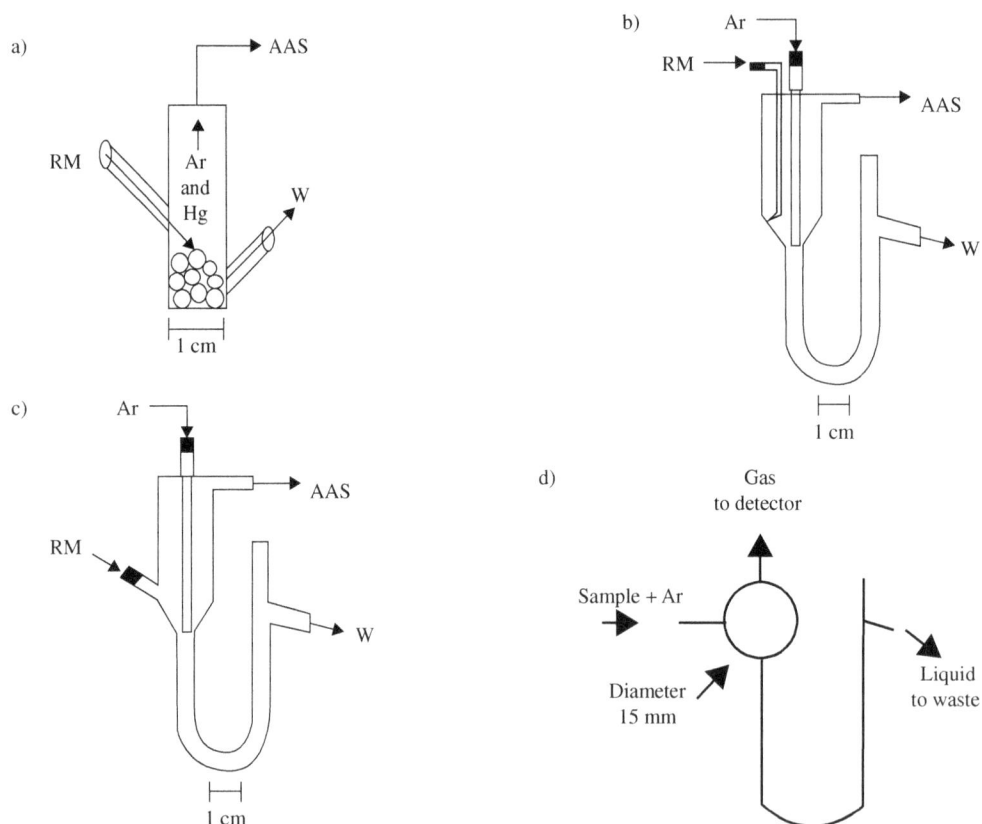

Figure 7.7. Examples of gas–liquid separators used in flow systems (not drawn to the same scale).
Source: a, b, and c—have been reproduced with permission from Hanna et al. (1993) © by Royal Society of Chemistry. d—is reprinted with permission from Chan (1985) © 2014 by American Chemical Society.

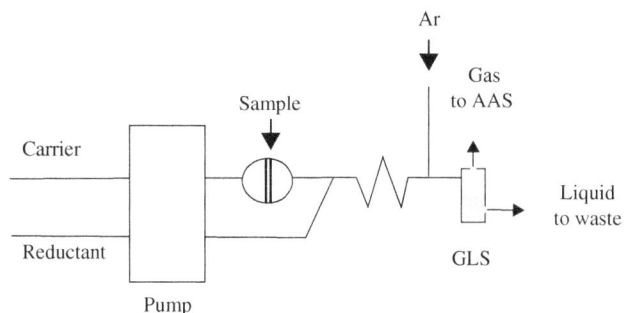

Figure 7.8. Basic flow-injection analysis (FIA) system for online gas–liquid separation.

7.6.2 Approaches for Preconcentration in the Gas Phase

Further increases in sensitivity can be obtained by further concentration of the analytes released to the gas phase, giving rise to very high analytical sensitivities mainly limited by blank values and some details were already discussed in Chapter 6 (Section 6.5). In this context, the amalgamation of atomic mercury with noble metals and posterior release by heating is a well-established method (commercial equipment is available) to further concentrate mercury vapor. The trap usually consists of gold-coated silica powder, gold-coated sand, or amalgams such as gold–platinum gauze.

Approaches to trap the hydrides involve the use of cryogenic traps or the retention of volatile hydrides on a slightly heated surface (normally graphite coated with noble metals) followed in both cases by release through further heating. Such approaches are most frequently used in connection with detection by ETAAS.

Continuously heated graphite tubes, wherein the volatile analyte hydride enters a hot furnace ($1,800–2,300°C$) and is atomized during its transit time through the device, have had limited use. However, trapping techniques that couple flow hydride generation with graphite furnace analyte collection are of interest because a clean, rapid separation/preconcentration of the analyte from the matrix is achieved in a simple way, and automated commercial instruments are available for such purposes. In such systems, the graphite furnace is used as both the hydride-trapping medium and the atomization cell. The hydride purged from the generator is trapped in the preheated furnace, usually at $300–600°C$, until the evolution of hydride is completed. The trapped analyte is subsequently atomized at temperatures generally greater than $2,000°C$. This technique has been shown to enhance the sensitivity significantly and to eliminate effectively the possible influence of the hydride-generation kinetics on the signal shape. The nature of the graphite tube is expected to affect greatly the efficiency of hydride adsorption; it has been shown that the coating of the graphite tube with Pd, Zr, Ag, or Pd–Ir mixtures, improves the sensitivity and precision significantly. In situ trapping of previously volatilized species and final ETAAS detection has been successfully applied to the determination of As, Hg, Ge, Bi, Pb, In, Sb, Se, Sn, Cd, or Te in a great variety of matrices.

7.7 MINIATURIZED PRECONCENTRATION METHODS BASED ON LIQUID–LIQUID EXTRACTION

Classical liquid–liquid extraction, based on transfer of analyte from the aqueous sample to a water-immiscible solvent, has been widely employed in the past for sample preparation. However, inconveniences such as emulsion formation, the use of large sample volumes, and toxic organic solvents have restricted its current use. Miniaturization of the liquid–liquid extraction technique can be achieved by a drastic reduction of the extractant phase volume and new methodologies have arisen such as single-drop microextraction (SDME), hollow-fiber liquid-phase microextraction, and dispersive liquid–liquid microextraction (Pena-Pereira et al. 2009).

In particular, SDME is a preconcentration technique based on the use of a microdrop of water-immiscible extractant phase suspended from the tip of a microsyringe needle and exposed to a stirred aqueous phase containing the sample (samples should be perfectly clean to enhance the stability of the drop at the tip of the needle). After extracting for a prescribed period of time, the drop is retracted into the microsyringe needle and finally injected into the detector to obtain the corresponding analytical signal. SDME is a simple, low-cost, and fast sample-preparation technique based on a great reduction of the extractant phase-to-sample volume ratio. SDME is not exhaustive, since only a small fraction of analyte(s) is extracted/preconcentrated for analysis.

SDME can be adapted to continuous-flow microextraction. In this case, the sample is pumped continuously at a constant flow rate and the extraction takes place within a glass extraction chamber. When the extraction chamber is full of sample, a drop is formed at the tip of a microsyringe needle (see Figure 7.9). The flow induces mass transfer in the drop via momentum transfer (the extraction phase indirectly experiences convection as a consequence of convection of the aqueous sample). The rate of extraction increases with increasing flow rate of the aqueous solution. The sample flow rate should ensure an effective microextraction of analytes without drop dislodgement or bubble formation. The selection of the extractant should be based on a comparison of selectivity, extraction efficiency, incidence of drop loss, rate of drop dissolution, and toxicity. A high boiling point reduces evaporative losses and a high surface tension increases the cohesive forces at the interface, hence reducing solvent solubilization.

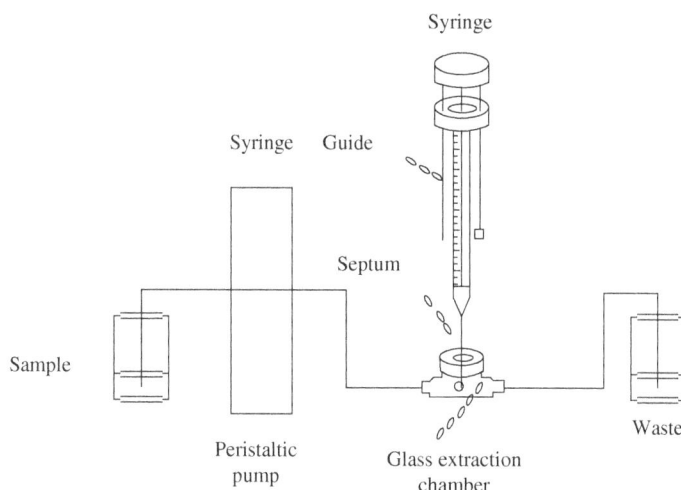

Figure 7.9. Schematic representation of continuous-flow microextraction.
Source: Reproduced with permission from Pena-Pereira et al. (2009) © by Elsevier.

For example, use (Cao et al. 2008) of a continuous-flow microextraction method for the determination of lead in water samples has been reported. In this case, a drop of 1-phenyl-3-methyl-4-benzoyl-5-pyrazolone in benzene was exposed to an aqueous sample pumped at a constant flow rate of 0.5 mL/min and the extract determined by ETAAS.

7.8 SAMPLE DIGESTION

The possibility to decompose/dissolve solid samples online with the atomic absorption spectrometers has attracted a great deal of interest. Although, some organo-species need to be first decomposed in order to achieve online volatile-species generation (such online digestion is particularly interesting in metal speciation, for species decomposition at the interface between the exit of a liquid chromatographic column and the hydride- or cold-vapor-generation systems).

Sample preparation and sample introduction in analytical systems are rapidly evolving toward greener, simpler, and more cost-effective methodologies. Implementation of ultraviolet (UV) irradiation can provide remarkable advantages in the area of trace-element analysis and speciation, since methods for removal of organic matter, decomposition of organic–metal complexes, and vapor generation can be driven by photo-oxidation or photoreduction processes with total or partial replacement of chemicals employed in conventional approaches.

On the other hand, sample heating in a microwave oven is not produced by an external source of heat but by direct interaction between the microwave radiation and the sample molecules. This eliminates the heat-conduction stage, making processes faster, more efficient, and requiring less energy than classical heating methods.

Thus, systems for online sample photo-oxidation or its online microwave-assisted digestion are briefly considered next.

7.8.1 Online Photo-Oxidation Flow Systems

The use of UV radiation (in some cases in combination with a strong oxidizing agent such as hydrogen peroxide or peroxodisulfate) to decompose organic matter is widely described in the literature. Furthermore, UV irradiation together with a chemical oxidant has been employed for degradation of metal species to facilitate the use of vapor-generation techniques in atomic spectrometry (Bendicho et al. 2010).

A mercury lamp is most commonly employed as the UV source, while most UV photoreactors are made of quartz or PTFE. In addition to its low UV absorption, a PTFE material offers favorable properties (such as lower cost, easier handling, and lower fragility) than quartz for constructing tubular photoreactors. In the area of flow systems for analytical purposes, several photoreactor designs have been proposed that are able to obtain appropriate efficiencies (Rubio et al. 1995). Usually, the lamp is wrapped with a coil of tubing (e.g., PTFE) through which the sample flows. To increase the light intensity reaching the coil and to prevent eye exposure to UV radiation, the unit is enclosed in aluminium foil.

7.8.2 Online Microwave-Assisted Digestion

A priori, microwave digestion in a flow was expected to be associated with serious problems derived from the vigorous chemical conditions, elevated temperatures, long digestion times frequently required to obtain complete decomposition, and high pressures (so, an additional problem to face is the evacuation of the gases produced during the digestion step). However, interesting instrumental devices have been developed trying to overcome these problems (Burguera et al. 2001).

The microwave oven can be incorporated into flow manifolds, which may be offline or online connected to the atomic detector. For example, in order to analyze samples requiring long digestion times, the flow can be interrupted for a period of time, while the sample is in the oven, resulting in stopped-flow digestion systems. In some designs, the oven is modified by placing an electric fan to vent hot air during operation and to help in cooling the tube at the end of the digestion. Frequently, a cooling system is connected online after the digestor. Also, in some designs, a backpressure regulator or a diffusion cell connected to a vacuum pump is located prior the spectrometer in order to remove the fumes produced during acid decomposition of organic materials.

Online microwave systems have been proposed for a variety of samples, such as botanical and animal tissues, urine, sediments, whole blood, and water. The samples can be injected into the flow manifold as slurries. It is interesting to note that studies on trace-element speciation have been undertaken in which a microwave oven is online located between the exit of a column for high-performance liquid chromatography (HPLC) and a cold-vapor or hydride-generation system. The oven decomposes online separated organo-species (e.g., of arsenic, mercury, or selenium) thus allowing an efficient online generation of the volatile derivatives (see Section 7.10.3) and final robust elemental detection by AAS.

7.9 CHROMATOGRAPHIC SEPARATIONS COUPLED ONLINE TO ATOMIC ABSORPTION SPECTROMETRY

Metal speciation concerns the identification and quantification of specific forms of a given metal. As different forms of an element may exhibit different toxicities and mobilities in the environment, it is clearly important to be able to distinguish between the individual species present in a given sample (Sanz-Medel 1998). In this context, total determinations of a toxic element by atomic spectroscopy are insufficient today and sometimes misleading if used to assess biological impact, for example, arsenobetaine is not toxic, methyl-mercury is much more toxic than inorganic mercury, and tributyltin is a most potent biocide, while Sn(IV) is not. Therefore, additional "speciation" information to complement total toxic element determinations is being increasingly demanded in the nutritional, toxicological, environmental, and clinical/biological fields.

Analytical approaches able to provide reliable molecular information of the sought-after particular species of the element under study are needed for speciation purposes. Of course, atomic methods are by definition non-speciating methods as the atomizer destroys the molecules. However, the so-called hybrid techniques, consisting in the coupling of a powerful separation (generally chromatographic techniques) with a sensitive element-specific atomic detector, offer a valuable combination. For discontinuous atomic detectors such as ETAAS, the separation can be carried out "off-line" with the detector. However, it is more convenient to couple online the separation unit to a continuous detector, such as flame atomic absorption spectrometry (FAAS), ICP-AES, or ICP-MS.

The selection of the separation technique depends on the nature and physical properties of the species of the element to be determined. When volatile, thermostable, neutral species (or able to produce them by chemical derivatization) have to be determined, the technique of choice is gas chromatography (GC). For example, Grignard reagents such as butylmagnesium chloride dissolved in THF are used to convert methyl- and inorganic mercury to volatile nonpolar butylmercury derivatives. Alternative derivatization reagents of great interest are sodium tetraalkylborates (such as sodium tetraethylborate) because they allow for carrying out the derivatization process in an aqueous medium.

For nonvolatile, thermally unstable, or charged compounds, the choice should be the more versatile but less-sensitive HPLC method. Provided that the target metal(loid) forms volatile derivatives, the exit of the column is online connected to a gas–liquid flow generator system. If a species of such a metal(loid) does not form a volatile derivative, approaches to decompose online such eluted species using oxidizing agents, UV radiation, or microwave energy prior to the gas-generation step have been successfully tested (see Section 7.10.3) before AAS measurements.

Finally, just a few words to comment that some promising work in chromatographic detection by AAS has been carried out using a diode laser as light source and flame or plasma as atomizers (Zybin 2004). Many HPLC methods employ high concentrations of organic solvents or salts and an important advantage of FAAS is that the flame is more robust against these compounds compared with a plasma (e.g., in ICP-MS), providing higher flexibility for appropriate chromatographic condition selection. The use of diode lasers for FAAS typically improves the sensitivity of the technique by one to three orders of magnitude while maintaining the advantages of low cost, robustness, and ease of use, thus providing an alternative to complicated and expensive detection systems for speciation analysis. Besides, elements such as Cl (837.82 nm), C (833.74 nm), and H (656.45 nm) can be easily monitored by AAS using such diode lasers. For this reason, diode laser AAS in combination with plasmas such as the microwave-induced plasma offer a high potential as a gas chromatographic detector of organohalides.

7.10 APPLICATIONS AND EXAMPLE CASE STUDIES

The following selected case studies illustrate a few of the many applications brought about by the combination of flow systems with AAS.

7.10.1 Online Aluminium Preconcentration and Its Application to the Determination of the Metal in Dialysis Concentrates

Certain clinical disorders that have been identified in renal-failure patients undergoing regular dialysis are associated with aluminium loading in the human body. For this reason, the solutions used for dialysis treatment should be checked for very low levels of Al. However, the determination of Al in concentrates for hemodialysis constitutes a formidable challenge since a low concentration of the metal must be determined in a dramatically high concentration of inorganic salts (e.g., about 10^5 mg/L of sodium ions and 10^3 mg/L of calcium ions).

This problem was successfully overcome some time ago by the use of an online minicolumn packed with Chelex 100. Figure 7.10 shows the flow manifold. The setup consisted of a peristaltic pump (A), a septum for sample injection (B), a mixing coil, a second valve for eluent injection (C), a minicolumn (D), and a third valve (E) placed after the minicolumn to minimize clogging of the nebulizer. As shown in Figure 7.9a, the sample injected through the septum mixes with the carrier along the mixing coil and the aluminium is

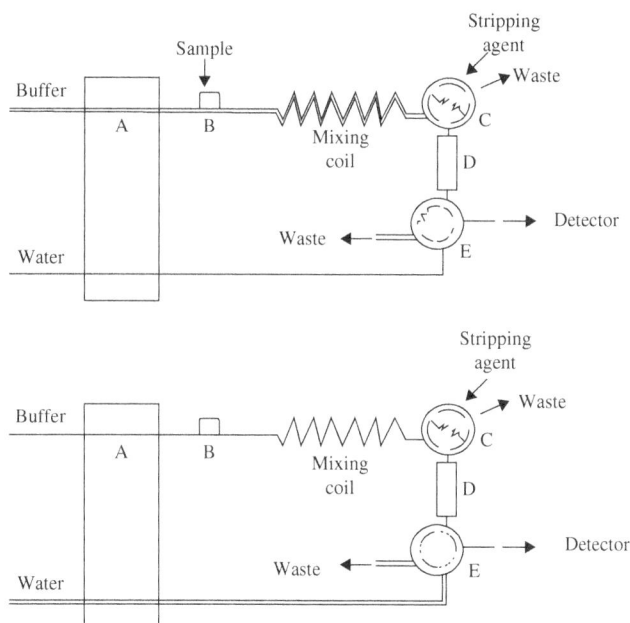

Figure 7.10. Flow diagram of the system used for the determination of Al. (a) preconcentration step; and (b) elution step. A, peristaltic pump; B, C, and E, injection valves; D, minicolumn. In each figure the solid black flow line corresponds to the channel entering the nebulizer.

Source: Reproduced with permission from Pereiro et al. (1990) © by Royal Society of Chemistry.

retained on the minicolumn. During this part of the cycle the nonretained matrix salts go to waste and water is continuously pumped to the detector. Figure 7.9b shows the step following once the preconcentration cycle is completed. First, valve E is opened releasing the solution (100 µL of 2M HCl) is inserted in the flow stream by turning valve C, thereby eluting Al directly into the nebulizer of the spectrometer (Pereiro et al. 1990).

This configuration not only prevents clogging of the nebulizer, but also allows for careful control of optimum experimental conditions for the determination of Al (during the preconcentration step, a standard solution of the metal can by pumped instead of water through the secondary line for detector control purposes). Using this manifold, a detection limit of 15 µg/L was obtained with FAAS detection for 1-mL sample injections. Of course, this detection limit can be improved by preconcentrating higher volumes of sample. The methodology was successfully tested for the determination of Al in dialysis concentrates.

7.10.2 Indirect Atomic Absorption Spectrometric Determination of Iodine in Milk Products

The use of flow-separation systems allows important increases of the analytical reproducibility achievable by indirect atomic spectrometry determinations. For example, the indirect determination of iodine in commercially available infant formula used as baby foods and powdered milks has been successfully achieved with a flow manifold online coupled to FAAS (Yebra and Bollaín 2010).

In the procedure, the sample or standard solutions containing iodine (as iodide) and a precipitating reagent (silver solution) are merged online and are mixed in the precipitation coil, where the precipitation occurs; the silver precipitate formed is retained on a filter device. Afterward, a washing solution (diluted ammonia solution) is introduced into the flow system. Finally, 0.3 mol/L thiosulfate flows through the filter to dissolve the silver iodide precipitate and, then, dissolved silver iodide is transported online to the nebulizer for silver detection. The obtained limit of detection was 2.7 µg/L, and the repeatability (expressed as relative standard deviation [%RSD]) checked on samples containing 2.55 µg/g iodine ($n = 11$) was 1.3%. The sample throughput achieved was ca. 17 samples/h.

7.10.3 High-Performance Liquid Chromatography——Microwave Digestion——Hydride Generation——AAS for Inorganic and Organic Arsenic Speciation in Fish Tissue

Arsenic is present in the environment in a number of different inorganic and organic forms, due to its participation in complex biological and chemical processes. As was described in Chapter 6 (Section 6.6.1), naturally occurring arsenic species in water and soil include As(III), As(V), monomethylarsonic acid (MMA), and dimethylarsinic acid (DMA). Other organoarsenicals such as arsenobetaine (AsB), arsenocholine (AsC), and arsenosugars may occur in biological tissues.

Arsenic species show different toxicities and chemical behaviors, leading to differences in their metabolism. Although complex organoarsenicals such as AsB and AsC are considered to be tolerated by living organisms, it has been reported that under specific conditions AsB present in marine biomass could be transformed into trimethylarsine, which is a toxic species, and therefore the determination of AsB and AsC in foods is important.

Some of the most suitable methods proposed in the literature to carry out arsenic speciation are based on the combination of HPLC with online detection by hydride generation coupled to an atomic spectrometer. The introduction of the hydride generation as a post-column-derivatization step leads to increased sensitivity and matrix removal, because the volatile hydrides are separated from the liquid residue. However, when hydride generation is employed for arsenic speciation it must be pointed out that some bio-organoarsenicals, such as AsB, AsC, and arsenosugars, are not detected by this approach, because they are resistant to acid digestion and do not form volatile hydrides. The use of online microwave digestion allows for fast decomposition of the species with great operation simplicity. Figure 7.11 depicts an HPLC–microwave–HG–AAS system for arsenic speciation in fish tissue ($K_2S_2O_8$ is used as oxidant).

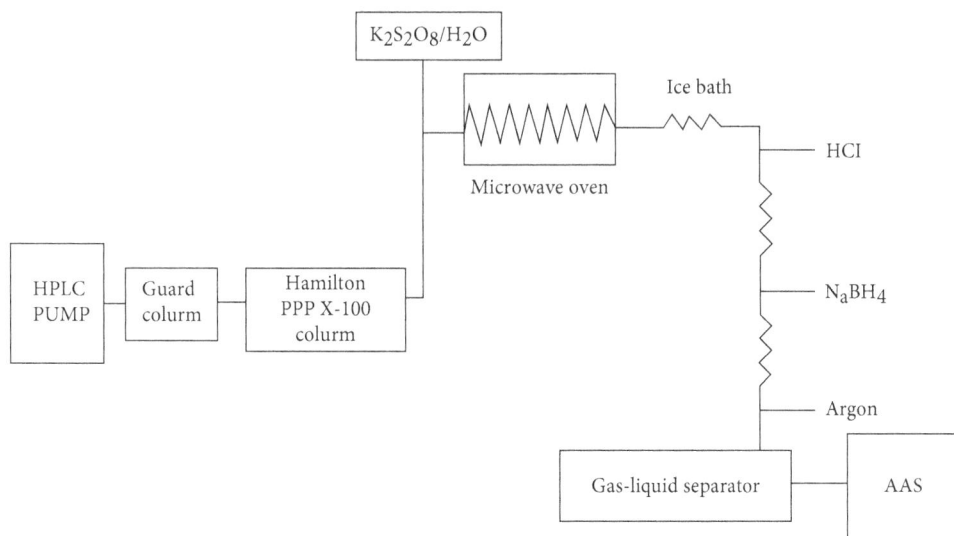

Figure 7.11. High-performance liquid chromatography (HPLC)–microwave–hydride-generation (HG)–atomic absorption spectroscopy (AAS) system for arsenic speciation.

Source: Reproduced with permission from Villa-Lojo et al. (2002) © by Elsevier.

It was observed than when HPLC is used for arsenic-species separation, AsB and As(III) peaks overlap under the chromatographic conditions used for most columns. To solve this problem, two separate injections, one using an oxidation step and the other without it, were proposed. Another alternative consists of previously isolating the AsB and AsC from the other arsenic species (As(III), As(V), MMA, and DMA) using a Sep-Pak® cartridge. The retained AsB and AsC are later eluted and injected into the chromatographic system.

8

Emerging Fields of Applications, Chemometrics, Quality-Control, and Troubleshooting

This final chapter includes a short description of some emerging fields of atomic absorption spectrometry applications, particularly the analysis of nanoparticles, analytical proteomics, and metallomics. Considering the present practical interest of using multivariate data analysis (chemometric) techniques in analytical atomic absorption spectrometry, a brief overview of such a synergic combination is also discussed here; this is followed by another practical discussion section dedicated to quality control guidelines and troubleshooting.

8.1 EMERGING FIELDS OF ATOMIC ABSORPTION SPECTROMETRY APPLICATIONS

Today, flame atomic absorption spectrometry (AAS) and Electrothermal atomic absorption spectrometry (ETAAS) are well-established analytical tools that are being routinely used for elemental analysis in environmental, clinical, food, and industrial applications. Although AAS is less sensitive than inductively coupled plasma (ICP) coupled to mass spectrometry (MS), AAS-based techniques are very useful (and, importantly, much less-expensive tools) for applications in several new fields of application, for example, studies related to metallomics and in nanoscience and nanotechnology. It is important to note as well that AAS is more tolerant to complex matrices than ICP-MS techniques. In addition, a sample amount of about 100 µL is sufficient to perform an ETAAS analysis, whereas at least a 10 times larger sample is required to ensure achieving the analytical performance of ICP.

Most studies in the biological field use AAS-based techniques to evaluate the total concentration of each biometal under study in a given sample. Good illustrations are the determination of liver iron content by flame AAS for investigation related with metalloproteomic analysis of hepatic iron overload in mice (Petrak et al. 2007), or the measurement of metal solution uptake of vegetals (or plant cellular models) by AAS via analyzing concentrations of the metals under study in the residual liquid (e.g., study of the influence of potassium supply in the metabolomic, proteomic, and biophysical studies of *Arabidopsis thaliana* cells exposed to a caesium stress [Le Lay et al. 2006]).

AAS has been also employed to study the evolution, the nature, and the stoichiometry of metallodrug–protein adducts (Timerbaev et al. 2011). Such analyses encounter fairly high concentrations of target metals, sufficiently so, that even flame AAS (FAAS) can be employed. In order to determine the albumin-binding characteristics of a set of cytotoxic mononuclear and dinuclear Pt(II) complexes (including cisplatin), the unbound fraction of protein was quantified by ETAAS after ultrafiltration and ethanolic precipitation. This treatment was used by the researchers to obtain, in addition to the unbound drug isolated by ultrafiltration, the reversible lipophilic bound drug suitable for AAS measurements.

Nanoscience and nanotechnology enjoy a bi-directional relationship with AAS. That is:

- In one way, AAS methods have been used for the characterization of nanomaterial synthesis (e.g., to measure the actual metal concentrations on the synthesized nanoparticles) as well as to study bioaccumulation and toxicity of metal(loid) nanoparticles in living organisms, after their administration. In such cases, levels in blood and in different organs are examined to investigate efficient uptake.
- On the other hand, nanomaterials have been exploited in order to improve the performance of AAS methods in recent years. With interesting properties such as a large surface area to volume ratio, high chemical and thermal stability, and excellent mechanical strength, nanomaterials can serve as efficient sorbents in sample preconcentration (Lum et al. 2014). Metal-oxide nanoparticles such as TiO_2 and Al_2O_3 have been tested for such purposes. Fe_3O_4 and alumina-coated Fe_3O_4 nanoparticles have also been employed by exploiting their magnetic properties (compared with nonmagnetic micro-absorbents, magnetic nanoparticles can lead to enhanced preconcentration efficiency due to the increased interactions between samples and sorbents) (Wang et al. 2012). Magnetic solid-phase extraction (MSPE) is more flexible and simpler than normal solid-phase extraction. It can be applied directly to bulk sample solutions; in such a case, an external magnetic field is used to immobilize the MSPE

sorbent for phase separation after extraction, thus saving the time of centrifugation. Carbon-containing nanomaterials have also been investigated for metal preconcentration, including single- and multiwalled carbon nanotubes and even graphene.

8.2 BASIC CHEMOMETRIC TECHNIQUES IN AAS

Chemometrics is a highly interfacial discipline which, using statistical and mathematical methods (such as multivariate statistics, applied mathematics, and computer science), is increasingly employed to help in solving problems in chemistry, biochemistry, medicine, biology, and chemical engineering. Chemometric techniques are particularly important in analytical chemistry in general, and in metabolomics and proteomics in particular. The number of publications reporting on the use of chemometric tools in atomic spectrometry is small compared with molecular spectrometry applications. However, chemometrics can help to implement robust atomic analytical methodologies by improving experimental design and optimization, or to achieve better analytical performance by resorting to multivariate calibration techniques such as principal components regression and partial least-squares methods (Andrade-Garda 2013).

With regard to optimization strategies, the simplex method and a modified simplex method (modification of the simplex algorithm allowing the size to be varied to adapt it to the experimental response) have been employed in many FAAS and ETAAS applications, as a powerful alternative to univariate optimization. Just to give two examples, the use of the simplex method for optimization of hydride generation parameters for direct determination of As(III) and total inorganic As by their in situ trapping with HG-ETAAS (Matusiewicz et Mroczkowska 2003), and the application of a modified simplex method for Cd separation and preconcentration in natural waters using a liquid membrane system with flame AAS detection (Granado-Castro et al. 2004), have both been reported.

Notwithstanding the fact that partial least-squares multivariate regressions have not been broadly applied in atomic spectrometry, this is by far the most common multivariate regression technique employed so far. Despite the relatively small number of reports, multivariate regression has proved to be very useful in those cases where analyte atomization is prone to complex interferences. For example, when slurries are analyzed by ETAAS sometimes spectral and chemical interferences cannot be totally avoided with the use of just chemical modifiers and optimized temperature programs. In such cases, the molecular absorption signal can be so high and structured that background correctors cannot be fully effective. In addition, the measurement at alternative wavelengths to avoid such problems may not provide a solution, due to low sensitivity for the trace levels to be measured. An illustrative application of partial least-squares regression is the mathematical modeling of complex interfering effects on the Sb AAS signal when soils, sediments, and fly ash samples were analyzed by ultrasonic slurry sampling-ETAAS (Felipe-Sotelo et al. 2006).

Finally, it is also interesting to note that artificial neural networks (ANNs) constitute a promising approach to cope with complex spectral problems in AAS. ANNs have interesting properties, for example, they work intrinsically with nonlinear functions, are flexible (in the sense that they do not require an underlying mathematical model), accept any relation between the spectra and the analyte(s)' concentration(s), are robust (i.e., they support small spectral deviations from the calibration data set), and so forth (Andrade-Garda 2013). In brief, ANNs offer special applicability to tackle problems that cannot be addressed by multivariate regression methods based on linear relationships (e.g., those involving spectral nonlinearities).

8.3 QUALITY-CONTROL GUIDELINES AND TROUBLESHOOTING

Fast production of reliable data with a minimum of errors is the goal of any routine analytical laboratory. To achieve this, an appropriate program for quality control must be implemented. Quality control in analytical chemistry refers to those processes and procedures designed to ensure that the results of laboratory analysis are consistent, comparable, accurate, reproducible, and within specified limits of precision. This implies that the errors (even those of a statistical nature that are unavoidable) have to be assessed in order to enable a decision to be made as to whether they are of an acceptable magnitude. Of course, unacceptable errors can also be discovered so that appropriate corrective actions can be taken.

All analytical laboratories are aware of the crucial importance of proper quality-assurance programs. A critical part is to have the standard operation procedures appropriately written and the analytical steps meticulously performed accordingly. Some relevant aspects include: calibration, use of blanks, performance characteristics of the procedure, and reporting of results. However, the written guidelines of a quality-control program are a necessary, but not a sufficient condition for obtaining and maintaining a proper statistical control. This is the role of quality assessment that includes internal methods coordinated within the laboratory (analysis of reference samples and control samples) and external methods organized and maintained by an external agency.

In the analysis of trace and ultratrace elements the possible sources of error are many. So, careful attention must be paid to all stages of the analytical procedure: sampling and pretreatment (e.g., homogenization and digestion), separations, analytical measurements (including quality of the chemical standards and performance of the instrument), and statistical data treatment. The quality-control procedures today implemented in many laboratories are helpful in keeping many sources of error under control; however, quality-control protocols should be continuously revisited, revised if necessary, in order to implement improved checking routines to secure the accuracy and precision aimed at.

Some useful tips to avoid errors, troubleshooting, and maintenance issues of instruments used for atomic absorption spectrometry are indicated below.

8.3.1 Flame AAS

Problems may include a misaligned burner slot, misaligned lamp, improper lamp current, unsuitable monochromator spectral bandpass, inappropriate flame stoichiometry, clogged nebulizer, improperly filled drain trap, among others. It is commonly considered that about 95% of the problems encountered are related to the light system, the nebulizer/burner, and the instrument cleanliness; the instrument's optics and electronics rarely fail. Most common problems accounting for absorbances lower than expected are related to the following.

8.3.1.1 Light system

Lamp alignment should be checked. Also, it could occur that wavelength is not tuned correctly or it is peaked on the wrong spectral line; here, special care should be paid to select the proper slit width (e.g., any element requiring a 0.2-nm slit will have more than one spectral line in the region). Vibrations in the laboratory can also cause wavelength drift; so, any vibrating equipment, including air compressors, should not be sited near the measurement instrument.

8.3.1.2 Nebulizer and Burner System

Analyzing solutions with suspended solids will sometimes lead to a blockage in the nebulizer hose or in the narrow inlet that it connects to (the sound coming from the sucking up of the solution will change and the absorbance reading will drop). In such a case, removal of the hose and careful unblocking of the inlet with a piece of fine wire is usually needed. Furthermore, it is important to note that about 90% of the nebulized solution goes to drain. Sometimes, the water does not drain properly, even if the bottle is not full. This will cause inconsistent results, due to poor uptake through the nebulizer to the flame. This situation requires clearing the blockage (usually an air bubble).

Special parameters to be controlled are the burner system alignment and the correct burner height. Also, the fuel-oxidation gas composition should be optimized and attention should be paid to the acetylene tank pressure. Acetylene tanks contain acetone (as acetylene-dissolving agent) for safety purposes: When the acetylene pressure drops, a decrease in absorbance and an increase in background, due to acetone, can be expected.

8.3.1.3 System Cleanliness

Smudges and chemical residues reduce light throughput and increase noise; therefore, lamp and sample compartment windows should be appropriately checked and, if necessary, cleaned, along with other factors such as deposits in the plastic capillary and in the nebulizer, which reduce sample uptake rate.

8.3.2 Electrothermal AAS

With regard to ETAAS, the most common problems arise from the light system, the autosampler, the furnace workhead, and the background correction system. As the problems with the light system should be similar to those for FAAS, described earlier, additionally, special attention should be paid to three other possible error sources.

8.3.2.1 Autosampler

Today autosamplers have the capacity to take up several separate volumes (e.g., of sample, standard, and diluent solutions) and mix them in the graphite furnace. It is usually taken for granted that the autosampler nominally delivered volumes are the actually delivered volumes. Physical parameters (e.g., density) can affect sample uptake and sample spreading on the graphite tube. Thus, the use of matrix-matched standards is recommended. A simple way that can help to check that the autosampler delivers the correct volumes is to inject equal volumes of sample and standard into the furnace, record the result, then reverse the sequence, inject standard and sample, and record the result. If different results are obtained, it can be assumed that "incorrect" volumes have been injected into the furnace.

It should be ensured that there are no bubbles in the dispensing syringe or deposits in the capillary tip. The autosampler should be properly rinsed (e.g., with 0.01% v/v HNO_3 plus a few drops of Triton X-100).

8.3.2.2 Furnace Workhead

The workhead position must be properly optimized (i.e., the light beam should pass through the center of the graphite furnace). Also, all steps of the atomizer temperature program should be optimized and a suitable matrix modifier selected. Again, although the graphite furnace correct temperatures for ashing or atomization are usually taken for granted from nominal values, in some cases (e.g., worn graphite cones or tubes) and on certain instruments, the actual temperature can differ by several hundred degrees from the set temperature. This can be devastating for certain trace element determinations. A simple way of monitoring this source of error is to visually check the temperature at which the graphite tube obtains a dull orange/red color. This normally occurs at approximately 800–850°C. If the deviation is more than approximately 50°C this is an indication that something is going wrong, and remedial actions should be taken (Jorhem 1995). Additionally, residues in the graphite tube may affect the sensitivity (e.g., they may cause contamination, higher noise, or higher background) and so an adequate conditioning of the tube before use is mandatory.

8.3.2.3 Background Correction

Nowadays all ETAAS instruments are equipped with facilities for background correction. However, in many cases, little attention is paid to whether the system actually corrects properly for background or not. For such a purpose, results for the analysis of two standards containing the same concentration of analyte but one of them also containing a potential spectral interference should be compared.

APPENDIX A

Buyer's Guide

A list of 10 manufacturers (alphabetic order) of atomic absorption spectrometers serving worldwide is collected below:

Agilent Technologies

5301, Stevens Creek Blvd,
Santa Clara, CA 95051, USA
Web: www.agilent.com

Analytik Jena AG

Konrad-Zuse-Straße 1,
07745, Jena, Germany
e-mail: info@analytik-jena.de
Web: www.analytik-jena.de

Aurora Instruments Ltd.

1001, East Pender Street,
Vancouver, BC, V6A 1W2 Canada
e-mail: info@aurora-instr.com
Web: www.aurora-instr.com

Buck Scientific, Inc.

58, Fort Point Street,
East Norwalk, CT 06855, USA
e-mail: sales@bucksci.com
Web: www.bucksci.com

GBC Scientific Equipment Pty Ltd.

2-4, Lakewood Blvd,
Braeside VIC 3195, Australia
e-mail: gbc@gbcsci.com
Web: www.gbcsci.com

Hitachi, Ltd.

6-6, Marunouchi 1-chome, Chiyoda-ku,
Tokyo 100-8280, Japan
Web: www.hitachi.com

Perkin Elmer, Inc.

Corporate Headquarters,
45, William Street,
Wellesley, MA 02481-4078, USA
e-mail: ProductInfo@perkinelmer.com
Web: www.perkinelmer.com

Shimadzu Corporation

Head Office,
1, Nishinokyo-Kuwabara-cho, Nakagyo-ku,
Kyoto 604-8511, Japan
Web: www.shimadzu.com

Teledyne Leeman Labs, Inc.

6, Wentworth Drive,
Hudson, NH 03051, USA
e-mail: SalesInfo@LeemanLabs.com
Web: www.teledyneleemanlabs.com

Thermo Fisher Scientific Inc.

81, Wyman Street,
Waltham, MA 02454, USA
Web: www.thermofisher.com

APPENDIX B

Glossary of Terms

Absorbance: the logarithmic function of the inverse of *transmittance*.

Absorptivity: also known as the molar extinction coefficient in molecular spectroscopy, it is the wavelength-dependent absorption of an analyte as a function of concentration and path length. It is expressed in units of concentration^{-1} cm^{-1}.

Analyte: a sample component whose concentration needs to be known.

Artificial neural networks: computational models inspired by animals' central nervous systems (in particular the brain) that are capable of machine learning and pattern recognition.

Ashing: or pyrolysis, is a step in an electrothermal atomic absorption spectrometry (ETAAS) program that is designed to remove matrix constituents that might interfere with the measurement of the analyte.

Atomization: the process of producing gaseous atoms. The atom-forming process usually requires a high temperature (except for cold-vapor methods), which is usually produced by a flame or by electrical current flowing through a resistive medium.

Atomization efficiency: efficiency to produce atomic vapor.

Background absorption: absorption of the source radiation by the flame itself and/or by concomitant species introduced into the atomizer.

Bandpass: see *Spectral bandpass.*

Bandwidth: see *Spectral bandwidth.*

Blank: an "ideal blank" contains all the sample constituents except the analyte. However, it is difficult to prepare an ideal blank because the concomitants and their concentrations are not usually known. In some cases, it is just sufficient to use the same solvent and any reagents employed to condition the sample.

Blank signal: includes the background signal due to optical signals from the sample container and the concomitants in the blank.

Beer–Lambert law: the law that defines a linear relationship between concentration and absorbance. It is often written as "absorbance = absorptivity path-length concentration."

Calibration: a procedure performed to relate the known concentration of standards containing the analyte to the detector signal.

Calibration curve: corresponds to the relationship of instrument response (absorbance) as a function of analyte concentration in the standards.

Chemical modifier: also known as "matrix modifier," is an element or compound that is added to the sample in ETAAS to increase the volatility of the matrix (and thus remove it during the ashing stage of the temperature program), or to decrease the volatility of the analyte element so that it can be atomized at higher temperatures.

Chemometrics: science of extracting information from chemical systems by multivariate data analysis–driven means.

Cold vapor: the chemical method by which atoms of mercury or cadmium are produced from a solution containing the corresponding ions.

Detection limit: this is indicative of the lowest amount of an analyte that can be detected with a specified degree of certainty. This is most often defined as three times the standard deviation of the blank measurement.

Deuterium background corrector: a strategy to correct for background absorption. This method employs a continuum radiation source (the deuterium lamp) that passes through the atomizer.

Diffraction grating: a plane or concave plate that is ruled with closely spaced grooves. The grating acts like a multislit source when collimated radiation strikes it. Different wavelengths are diffracted and constructively interfere at different angles.

Drain trap: an exit at the bottom of the mixing chamber that leads through a tube to a water-filled trap, allowing waste sample solution to drain from the mixing chamber but not permitting combustion gases to escape.

Dynamic range: sometimes known as "linear dynamic range" or "linear range," is the analyte concentration range over which response is a well-defined (usually linear) function of the analyte concentration.

Electrodeless discharge lamp: a radiation source for atomic absorption spectrometry that consists of a sealed quartz tube containing a small amount of the element of interest and an inert gas. The lamp is placed in a radiofrequency field, which excites the atoms to emit intense line radiation.

Flow spoiler: a plastic, fan-shaped device placed in the mixing chamber of a flame atomic absorption spectrometry (FAAS) instrument to improve the mixing of combustion gases with analyte solution droplets and facilitate the removal of large droplets down the drain trap at the bottom of the mixing chamber.

Focal plane: a plane where the spectrum is dispersed. The plain contains an aperture(s) or a slit(s), with the photodetector(s) located at the other side of the aperture(s) or slit(s).

Graphite tube: the most common atomization cell used in an electrothermal atomizer for AA. Typically made of *pyrolytic graphite* and bathed in an inert gas (most commonly argon) to prevent decomposition.

Ground state: the lowest energy state of an atom or molecule.

Half-width: also called "full width at half maximum" (FWHM), is the width in wavelength units at half the net peak height.

Hollow cathode lamp: the most common radiation source for atomic absorption spectrometry. It consists of a low-pressure inert-gas-filled tube containing an anode and a hollow cathode made from the element for which the lamp is to produce atomic line radiation.

Hydride generation: the method by which volatile hydride-forming elements are released from solutions containing their ions, commonly using sodium borohydride. The formed volatile hydrides are swept from solution by an inert gas and decomposed to atoms in a heated absorption cell.

Flow injection analysis (FIA): an automated method of (bio)chemical analysis in which a plug of sample is injected into a continuous flow of a carrier solution that mixes with other continuously flowing solutions before reaching a detector.

Flow rate (gas): the volumetric flow rate (mL/min) of combustible gases into the mixing chamber of a FAAS instrument, or of inert gas used for ETAAS and for chemical vapor generation methods.

Flow rate (solution): the volumetric flow rate (mL/min) of solution uptake into the nebulizer of a FAAS instrument.

Integration: a process for identifying and calculating the amount of a component by measuring the area greater than the baseline defined by the instrument blank over a specific time period. In conventional FAAS, integration times of a few seconds are most commonly used, since continuous signals are measured. In ETAAS, flow injection, and those gas-generation methods producing transient signals, the "*peak*" produced by the analyte is integrated from baseline to baseline.

Interference: is a substance present in the analytical sample or in the atomizer that adversely affects the magnitude of spectral signal measured for the analyte.

Linear dispersion: relates to how far apart in distance are two close wavelengths separated in the focal plane of a dispersive wavelength selector. Linear dispersion is conveniently expressed in mm/nm.

Matrix: components present in the sample that are not analyte(s).

Memory effects: influence exerted by the previous measured sample (or standard) in the signal generated by a given sample (or standard).

Metal speciation: identification and quantification of specific chemical forms of a given metal.

Metallome: comprehensive analysis of the entirety of metal and metalloid species within a cell or tissue type.

Metallomics: the study of metallome.

Mixing chamber: a chamber in which combustible gases are mixed with the sample solution droplets from the nebulizer and then transported toward the flame. Larger droplets (approximately 95% of the sample) are removed from the mixing chamber through the drain trap.

Modulation: the periodic variation of the radiation from the light source, most commonly electronically (and in former times mechanically with a chopper). This allows for discriminating against other sources of radiation that might reach the detector affecting the absorbance measurement corresponding to the analyte.

Monochromator: a wavelength selector used in optical spectrometers to isolate the radiation analytical useful from other radiation.

Nanoparticle: particles with diameters between 1 and 100 nanometers.

Nebulizer: component of the sample introduction system that draws aqueous solution into the premix chamber and converts it to a fine mist of small droplets that are swept into the flame.

Parts per billion or ppb: in a liquid corresponds to micrograms of analyte per liter of sample (μg/L) and in a solid to nanograms of analyte per gram of sample (ng/g).

Parts per million or ppm: in a liquid corresponds to milligrams of analyte per liter of sample (mg/L) and in a solid to micrograms of analyte per gram of sample (μg/g).

Peak: the transient signal increase, obtained using flow injection procedures, discontinuous gas-generation methods, or electrothermal atomizers, the height of which and area are related to the concentration of the analyte in a sample.

Photomultiplier tube: the most often used detector in an atomic absorption spectrometer. It consists of a vacuum tube containing an alkali-element photocathode that produces electrons when struck by photons of sufficient energy (the photoelectric effect). Each photoelectron is then multiplied by consecutive collisions with a series of dynodes so that the electrical signal produced by each incident photon is greatly amplified.

Platform atomization: also known as the L'vov platform, it is a small platform onto which a sample is placed inside the graphite furnace tube rather than placing the sample on the tube wall. This delays sample atomization until the gas temperature inside the furnace is higher than it would be for wall atomization, which reduces some interference effects.

Principal component analysis: a statistical procedure that uses orthogonal transformation to convert a set of observations of possibly correlated variables into a set of values of linearly uncorrelated variables called principal components. This transformation is defined in such a way that the first principal component has the largest possible variance and each succeeding component in turn has the highest variance possible under the constraint that it is orthogonal to the preceding components.

Pyrolytic graphite: a form of graphite manufactured by decomposition of methane gas at low pressure (about 1 torr) and heated to 2,000°C. The result is a product that is very highly ordered, chemically inert, impermeable, highly pure, and stable upto 3,000°C.

Qualitative analysis: the determination of the identity of the components in the sample.

Quantitative analysis: the determination of the amount or concentration of one/several component(s) of the sample.

Quality assessment: systematic evaluation of a service or how good an analysis is to determine its performance in relation to set standards.

Quality control: term used to describe the practical steps undertaken to ensure that errors in the analytical data are of a magnitude appropriate for the use to which the data will be utilized.

Refractory elements in AAS: elements that tend to form heat-stable compounds in the flame.

Resonance transition: a transition to or from the ground electronic level.

Resonance line: the resulting spectral line from a resonance transition.

Reciprocal linear dispersion: represents the number of wavelength intervals (e.g., nm) contained in each interval of distance (e.g., mm) along the focal plane.

Sample throughput: is the number of samples that can be analyzed, or elements that can be determined, per unit time.

Smith–Hieftje background corrector: a strategy to correct for background absorption in AAS that pulses the hollow cathode lamp at low and then at high current. During the high-current pulse, a large cloud of atoms is formed in front of the lamp cathode; this cloud prevents hollow cathode radiation from reaching the analyte in the atomizer. Thus, it allows discrimination between atomic absorption and other sources of absorption.

Spectral bandwidth: or spectral bandpass, is the half-width of the wavelength distribution passing by the exit slit of a dispersive wavelength selector.

Spectrometry: quantitative measurement of the intensity of electromagnetic radiation at one or more wavelengths with a photoelectric detector.

Spectrum: in optical spectroscopy corresponds to the display of the intensity of radiation emitted, absorbed, or scattered by a sample versus a quantity related to photon energy, such as wavelength or frequency.

Sputtering: process in which atoms or ions are ejected from a surface by the impact of charged particles.

Standard: solution or solid with a known concentration of analyte (used for instrument/method calibration).

Standard addition: a calibration method that compensates for matrix-induced enhancement or suppression of analyte signals. A known concentration of an analyte element is added to the sample and the instrument response to the known concentration of added element is used to calibrate the instrument response for the sample.

Sensitivity: slope of the linear plot of "instrument response versus analyte concentration." Traditionally, in AAS, the sensitivity is defined as the concentration of analyte that produces an instrument response of 0.0044 absorbance units (1% absorption).

Transmittance: the fraction of the incident electromagnetic radiation that is transmitted by a sample.

Troubleshooting: is a logical, systematic search for the source of a problem so that it can be solved and the product or process can be made operational again.

Validated method: a method that meets or exceeds certain sampling and measurement performance criteria.

Zeeman background corrector: strategy to correct for background absorption in atomic absorption that uses a magnetic field around the atomizer.

APPENDIX C

Standards

C.1 BRITISH STANDARDS INSTITUTION

In many instances, British Standards (BS), www.bsigroup.com, are automatically incorporated with standards from the International Standards Organization (ISO), www.iso.org. Besides, the European Committee for Standardization (www.cen.eu) and the European Committee for Electrotechnical Standardization (www.cenelec.eu) elaborate or adopt technical standards to promote free trading, the safety of workers and consumers, interoperability of networks, environmental protection, exploitation of research and development programs, and public procurement. EN corresponds to the acronym of "European Norms." Whenever a BS norm is a transposition of an EN, this should be indicated as follows:

BS EN 1811: 1999: Reference test method for release of nickel from products intended to come into direct and prolonged contact with the skin [12 pp].

For the case that an adopted EN norm proceeds from an ISO norm, this should be indicated as follows:

BS EN ISO 6869: 2001: Animal feeding stuffs. Determination of the contents of calcium, copper, iron, magnesium, manganese, potassium, sodium, and zinc. Method using atomic absorption spectrometry [26 pp].

Given below are five collected examples of BS standards not originating from EN or ISO standards.

BS 2000-455: 2000: Methods of test for petroleum and its products. Determination of manganese in gasoline. Atomic absorption spectrometry (AAS) method [4 pp].

BS 5766-6: 1984: Methods for analysis of animal feeding stuffs. Determination of calcium by atomic absorption spectrometry [8 pp].

BS 6075-11: 1981: Methods of sampling and test for sodium hydroxide for industrial use. Determination of mercury content (flameless atomic absorption method) [10 pp].

BS 7164-28.1: 1977: Chemical tests for raw and vulcanized rubber. Methods for determination of copper content. Atomic absorption spectrometry [14 pp].

BS 7317-3: 1990: Methods for analysis of high-purity copper Cu-CATH-1. Method for determination of antimony, arsenic, bismuth, selenium, tellurium and tin by hydride generation, and atomic absorption spectrometry [12 pp].

C.2 INTERNATIONAL STANDARDS ORGANIZATION

There is a wide variety of ISO Standards on atomic absorption spectrometry for analysis of elements in food, environment, raw materials, industrial products, and so forth. Below, a small selection of methods is listed that use flame atomic absorption spectrometry (FAAS), electrothermal atomic absorption spectrometry (ETAAS), or generation of hydride-AAS or cold vapor-AAS:

ISO 4744: 1984: Copper and copper alloys; determination of chromium content; Flame atomic absorption spectrometric method [3 pp].

ISO 5373: 1981: Condensed phosphates for industrial use (including food-stuffs); determination of calcium content; Flame atomic absorption spectrometric method [4 pp].

ISO 5889: 1983: Manganese ores and concentrates; determination of aluminium, copper, lead, and zinc contents; Flame atomic absorption spectrometric method [4 pp].

ISO 6101-6: 2011: Rubber—Determination of metal content by atomic absorption spectrometry—Part 6: Determination of magnesium content [8 pp].

ISO 6636-2: 1981: Fruits, vegetables, and derived products—Determination of zinc content—Part 2: Atomic absorption spectrometric method [4 pp].

ISO 7530-1: 1990: Nickel alloys; flame atomic absorption spectrometric analysis; Part 1: General requirements and sample dissolution [7 pp].

ISO 7530-8: 1992: Nickel alloys; flame atomic atomic absorption spectrometric analysis; Part 8: Determination of silicon content [3 pp].

ISO 7627-4: 1983: Hard metals; chemical analysis by flame atomic absorption spectrometry; Part 4: Determination of molybdenum, titanium, and vanadium contents from 0.01% to 0.5 % (m/m) [2 pp].

ISO 8294: 1994: Animal and vegetable fats and oils—determination of copper, iron, and nickel contents—graphite furnace atomic absorption method [6 pp].

ISO 8658: 1997: Carbonaceous materials for use in the production of aluminium—green and calcined coke—determination of trace elements by flame atomic absorption spectroscopy [8 pp].

ISO 9174: 1998: Water quality—determination of chromium—atomic absorption spectrometric methods [10 pp].

ISO 9668: 1990: Pulps; determination of magnesium content; flame atomic absorption spectrometric method [3 pp].

ISO 9965: 1993: Water quality; determination of selenium; atomic absorption spectrometric method (hydride technique) [5 pp].

ISO 10136-3: 1993: Glass and glassware; analysis of extract solutions; Part 3: Determination of calcium oxide and magnesium oxide by flame atomic absorption spectrometry [6 pp].

ISO 10775: 2013: Paper, board, and pulps—determination of cadmium content—atomic absorption spectrometric method [6 pp].

ISO 11047: 1998: Soil quality—determination of cadmium, chromium, cobalt, copper, lead, manganese, nickel, and zinc in aqua regia extracts of soil—flame and electrothermal atomic absorption spectrometric methods [18 pp].

ISO 11438-1: 1993: Ferronickel; determination of trace-element content by electrothermal atomic absorption spectrometric method; Part 1: General requirements and sample dissolution [5 pp].

ISO 11438-8: 1993: Ferronickel; determination of trace-element content by electrothermal atomic absorption spectrometric method; Part 8: Determination of indium content [3 pp].

ISO 11813: 2010: Milk and milk products—determination of zinc content—flame atomic absorption spectrometric method (2nd Edition) [6 pp].

ISO 12193: 2004: Animal and vegetable fats and oils—Determination of lead by direct graphite furnace atomic absorption spectroscopy [7 pp].

ISO 12740: 1998: Lead sulfide concentrates—determination of silver and gold contents—fire assay and flame atomic absorption spectrometric method using scorification and cupellation [26 pp].

ISO 12846: 2012: Water quality—determination of mercury—method using atomic absorption spectrometry (AAS) with and without enrichment [15 pp].

ISO 14377: 2002: Canned evaporated milk—Determination of tin content—Method using graphite furnace atomic absorption spectrometry [8 pp].

ISO 15248: 1998: Zinc sulfide concentrates—determination of silver and gold contents—fire assay and flame atomic absorption spectrometric method using scorification or cupellation [25 pp].

ISO 15774: 2000: Animal and vegetable fats and oils—determination of cadmium content by direct graphite furnace atomic absorption spectrometry [6 pp].

ISO 17191: 2004: Urine-absorbing aids for incontinence—measurement of airborne respirable polyacrylate superabsorbent materials—determination of dust in collection cassettes by sodium atomic absorption spectrometry [9 pp].

ISO 17239: 2004: Fruits, vegetables, and derived products—determination of arsenic content—method using hydride generation atomic absorption spectrometry [10 pp].

ISO 17378-2: 2014: Water quality—determination of arsenic and antimony—Part 2: Method using hydride generation atomic absorption spectrometry (HG-AAS) [22 pp].

ISO 17733: 2004: Workplace air—determination of mercury and inorganic mercury compounds—method by cold vapor atomic absorption spectrometry or atomic fluorescence spectrometry [51 pp].

ISO 17992: 2013: Iron ores—determination of arsenic content—hydride generation atomic absorption spectrometric method [17 pp].

References

Aastroem, O. February 1982. "Flow Injection Analysis for the Determination of Bismuth by Atomic Absorption Spectrometry with Hydride Generation." *Analytical Chemistry* 54, no. 2, pp. 190–193. Doi: http://dx.doi.org/10.1021/ac00239a011.

Alexiu, V.; and L. Vladescu. 2005. "Optimization of a Chemical Modifier in the Determination of Selenium by Graphite Furnace Atomic Absorption Spectrometry and Its Application to Wheat and Wheat Flour Analysis." *Analytical Sciences* 21, no. 2, pp. 137–141. Doi: http://dx.doi.org/10.2116/analsci.21.137.

Alonso, E.V.; A. García de Torres; and J.M. Cano Pavón. August 2001. "Flow Injection On-Line Electrothermal Atomic Absorption Spectrometry Review. *Talanta* 55, no. 2, pp. 219–232. Doi: http://dx.doi.org/10.1016/s0039-9140(01)00371-x.

Anawar, H.M. January 2012. "Arsenic Speciation in Environmental Samples by Hydride Generation and Electrothermal Atomic Absorption Spectrometry Review." *Talanta* 88, pp. 30–42. Doi: http://dx.doi.org/10.1016/j.talanta.2011.11.068.

Andrade-Garda, J.M., Ed. 2013. *Basic Chemometric Techniques in Atomic Spectroscopy.*, p. 300. 2nd revised Ed. Cambridge, UK: The Royal Society of Chemistry.

Arbab-Zavar, M.H.; M.M. Chamsaz; A.A. Youssefi; and M. Aliakbari M. August 2005. "Mechanistic Aspects of Electrochemical Hydride Generation for Cadmium." *Analytica Chimica Acta* 576, no. 2, pp. 215–220. Doi: http://dx.doi.org/10.1016/j.aca.2006.06.015.

Becker-Ross, H.; M. Okruss; S. Florek; U. Heitmann; and M.D. Huang. October 2002. "Echelle-Spectrograph as a Tool for Studies of Structured Background in Flame Atomic Absorption Spectrometry." *Spectrochimica Acta Part B: Atomic Spectroscopy* 57, no. 10, pp. 1493–1504. Doi: http://dx.doi.org/10.1016/s0584-8547(02)00107-6.

Bechlin, M.A.; J.A.G. Neto; and J.A. Nóbrega. July 2013. "Evaluation of Lines of Boron, Phosphorus, and Sulfur by High-Resolution Continuum Source Flame Atomic Absorption Spectrometry for Plant Analysis." *Microchemical Journal* 109, pp. 134–138. Doi: http://dx.doi.org/10.1016/j.microc.2012.03.013.

Bendicho, C.; F. Pena; M. Costas; S. Gil; and I. Lavilla. July 2010. "Photochemistry-Based Sample Treatments as Greener Approaches for Trace-Element Analysis and Speciation." *TrAC Trends in Analytical Chemistry* 29, no. 7, pp. 681–691. Doi: http://dx.doi.org/10.1016/j.trac.2010.05.003.

Bolea, E.; D. Arroyo; F. Laborda; and J.R. Castillo. May 2006. "Determination of Antimony by Electrochemical Hydride Generation Atomic Absorption Spectrometry in Samples with High Iron Content Using Chelating Resins as On-Line Removal System." *Analytica Chimica Acta* 569, no. 1–2, pp. 227–233. Doi: http://dx.doi.org/10.1016/j.aca.2006.03.076.

Burguera, J.L.; and M. Burguera. October 2001. "Flow Injection-Electrothermal Atomic Absorption Spectrometry Configurations: Recent Developments and Trends." *Spectrochimica Acta Part B: Atomic Spectroscopy* 56, no. 10, pp. 1801–1829. Doi: http://dx.doi.org/10.1016/s0584-8547(01)00338-x.

Burguera, J.L.; and M. Burguera. December 2004. "Analytical Applications of Organized Assemblies for On-Line Spectrometric Determinations: Present and Future." *Talanta* 64, no. 5, pp. 1099–1108. Doi: http://dx.doi.org/10.1016/j.talanta.2004.02.046.

Cal-Prieto, M.J.; M. Felipe-Sotelo; A. Carlosena; J.M. Andrade; P. López-Mahía; S. Muniategui; and D. Prada. January 2002. "Slurry Sampling for Direct Analysis of Solid Materials by Electrothermal Atomic Absorption Spectrometry. A Literature Review from 1990 to 2000." *Talanta* 56, no. 1, pp. 1–51. Doi: http://dx.doi.org/10.1016/s0039-9140(01)00543-4.

Camel, V. July 2003. "Solid Phase Extraction of Trace Elements. Review." *Spectrochimica Acta Part B: Atomic Spectroscopy* 58, no. 7, pp. 1177–1233. Doi: http://dx.doi.org/10.1016/s0584-8547(03)00072-7.

Cao, J.; P. Liang; and R. Liu. April 2008. "Determination of Trace Lead in Water Samples by Continuous Flow Microextraction Combined with Graphite Furnace Atomic Absorption Spectrometry." *Journal of Hazardous Materials* 152, no. 3, pp. 910–914. Doi: http://dx.doi.org/10.1016/j.jhazmat.2007.07.064.

Chan, C.C.Y. June 1985. "Semiautomated Method for Determination of Selenium in Geological Materials Using a Flow Injection Analysis Technique." *Analytical Chemistry* 57, no. 7, pp. 1482–1485. Doi: http://dx.doi.org/10.1021/ac00284a072.

D'Ulivo, A. June 2004. "Chemical Vapor Generation by Tetrahydroborate(III) and Other Borane Complexes in Aqueous Media. A Critical Discussion of Fundamental Processes and Mechanisms Involved in Reagent Decomposition and Hydride Formation." *Spectrochimica Acta Part B: Atomic Spectroscopy* 59, no. 6, pp. 793–825. Doi: http://dx.doi.org/10.1016/j.sab.2004.04.001.

D'Ulivo, A.; J. Dědina; Z. Mester; R.E. Sturgeon; Q. Wang; and B. Welz. 2011. "Mechanisms of Chemical Generation of Volatile Hydrides for Trace Element Determination (IUPAC Technical Report). *Pure and Applied Chemistry* 83, no. 6, pp. 1283–1340. Doi: http://dx.doi.org/10.1351/pac-rep-09-10-03.

Economou, A. May 2005. "Sequential-Injection Analysis (SIA): A Useful Tool for On-Line—Sample Handling and Pre-Treatment. *TrAC Trends in Analytical Chemistry* 24, no. 5, pp. 416–425. Doi: http://dx.doi.org/10.1016/j.trac.2004.12.004.

Emig, M.; R.I. Billmers; K.G. Owens; N.P. Cernansky; D.L. Miller; and F.A. Narducci. July 2002. "Sensitive and Selective Detection of Paramagnetic Species Using Cavity Enhanced Magneto-Optic Rotation." *Applied Spectroscopy* 56, no. 7, pp. 863–868. Doi: http://dx.doi.org/10.1366/000370202760171536.

Evans, E.H.; M. Horstwood; J. Pisonero; and C.M.M. Smith. 2013. "Atomic Spectrometry Update: Review of Advances in Atomic Spectrometry and Related Techniques." *Journal of Analytical Atomic Spectrometry* 28, no. 6, pp. 779–800. Doi: http://dx.doi.org/10.1039/c3ja90029k.

Felipe-Sotelo, M.; M.J. Cal-Prieto; M.P. Gómez Carracedo; J.M. Andrade; A. Carlosena; and D. Prada. July 2006. "Handling Complex Effects in Slurry-Sampling-Electrothermal Atomic Absorption Spectrometry by Multivariate Calibration." *Analytica Chimica Acta* 571, no. 2, pp. 315–323. Doi: http://dx.doi.org/10.1016/j.aca.2006.05.004.

Fernandez de la Campa, M.R.; E. Segovia Garcia; M.C. Valdes-Hevia y Temprano; B. Aizpun Fernandez; J.M. MarchanteGayon; and A. Sanz-Medel. June 1995. "Effects of Organised Media on the Generation of Volatile Species for Atomic Spectrometry." *Spectrochimica Acta Part B: Atomic Spectroscopy* 50, no. 4–7, pp. 377–391. Doi: http://dx.doi.org/10.1016/0584-8547(94)00155-o.

Frech, W. June 1996. "Recent Developments in Atomizers for Electrothermal Atomic Absorption Spectrometry." *Analytical and Bioanalytical Chemistry* 355, no. 5–6, pp. 475–486. Doi: http://dx.doi.org/10.1007/s0021663550475.

Granado Castro, M.D.; M.D. Galindo-Riaño; and M. García-Vargas. April 2004. "Separation and Preconcentration of Cadmium Ions in Natural Water Using a Liquid Membrane System with 2-Acetylpyridine

Benzoylhydrazone as Carrier by Flame Atomic Absorption Spectrometry." *Spectrochimica Acta Part B: Atomic Spectroscopy* 59, no. 4, pp. 577–583. Doi: http://dx.doi.org/10.1016/j.sab.2004.01.006.

Guo, X.W.; and X.M. Guo. 1995. "Determination of Cadmium at Ultratrace Levels by Cold Vapour Atomic Absorption Spectrometry." *Journal of Analytical Atomic Spectrometry* 10, no. 11, pp. 987–991. Doi: http://dx.doi.org/10.1039/ja9951000987.

Hanna, C.P.; P.E. Haigh,; J.F. Tyson; and S. McIntosh. 1993. "Examination of Separation Efficiencies of Mercury Vapour for Different Gas-Liquid Separators in Flow Injection Cold Vapour Atomic Absorption Spectrometry with Amalgam Preconcentration." *Journal of Analytical Atomic Spectrometry* 8, no. 4, pp. 585–590. Doi: http://dx.doi.org/10.1039/ja9930800585.

Heitmann, U.; M. Schütz; H. Becker-Ross; and S. Florek. July 1996. "Measurements of the Zeeman-Splitting of Analytical Lines by Means of a Continuum Source Graphite Furnace Atomic Absorption Spectrometer with a Linear Charge Coupled Device Array." *Spectrochimica Acta Part B: Atomic Spectroscopy* 51, no. 9–10, pp. 1095–1105. Doi: http://dx.doi.org/10.1016/0584-8547(96)01504-2.

Hieftje, G.M. 1989. "Atomic Absorption Spectrometry—Has It Gone or Where Is It Going? *Journal of Analytical Atomic Spectrometry* 4, no. 2, pp. 117–122. Doi: http://dx.doi.org/10.1039/ja9890400117.

Inczédy, J.; T. Lengyel; and A.M. Ure. 1998. *International Union of Pure and Applied Chemistry. Compendium of Analytical Nomenclature. Definitive rules 1997.*, p. 964. Oxford, England: Blackwell Science.

Jorhem, L. September 1995. "Dry Ashing, Sources of Error, and Performance Evaluation in AAS." *Mikrochimica Acta* 119, no. 3–4, pp. 211–218. Doi: http://dx.doi.org/10.1007/bf01244000.

Le Lay, P.; M.P. Isaure; J.E. Sarry; L. Kuhn; B. Fayard; J.L. Le Bail; O. Bastien; J. Garin; C. Roby; and J. Bourguignon. November 2006. "Metabolomic, Proteomic and Biophysical Analyses of Arabidopsis Thaliana Cells Exposed to a Caesium Stress. Influence of Potassium Supply." *Biochimie* 88, no. 11, 1533–1547. Doi: http://dx.doi.org/10.1016/j.biochi.2006.03.013.

Lenehan, C.E.; N.W. Barnett; and S.W. Lewis. August 2002. "Sequential Injection Analysis." *The Analyst* 127, no. 8, pp. 997–1020. Doi: http://dx.doi.org/10.1039/b106791p.

Lum, T.S.; Y.K. Tsoi; and K.S.Y. Leung. 2014. "Current Developments in Clinical Sample Preconcentration Prior to Elemental Analysis by Atomic Spectrometry: A Comprehensive Literature Review." *Journal of Analytical Atomic Spectrometry* 29, no. 2, pp. 234–241. Doi: http://dx.doi.org/10.1039/c3ja50316j.

Matusiewicz, H.; and M. Mroczkowska. 2003. "Hydride Generation from Slurry Samples After Ultrasonication and Ozonation for the Direct Determination of Trace Amounts of As (III) and Total Inorganic Arsenic by Their In Situ Trapping Followed by Graphite Furnace Atomic Absorption Spectrometry." *Journal of Analytical Atomic Spectrometry* 18, no. 7, pp. 751–761. Doi: http://dx.doi.org/10.1039/b302217j.

Miró, M.; and E.H. Hansen. June 2013. "On-Line Sample Processing Involving Microextraction Techniques as a Front-End to Atomic Spectrometric Detection for Trace Metal Assays: A Review." *Analytica Chimica Acta* 782, pp. 1–11. Doi: http://dx.doi.org/10.1016/j.aca.2013.03.019.

Montero, R.; M. Gallego; and M. Valcárcel. 1988. "Indirect Atomic Absorption Spectrometric Determination of Sulphonamides in Pharmaceutical Preparations and Urine by Continuous Precipitation." *Journal of Analytical Atomic Spectrometry* 3, no. 5, pp. 725–729. Doi: http://dx.doi.org/10.1039/ja9880300725.

Nakadi, F.V.; L.R. Rosa; and M.A.M.S. da Veiga. October 2013. "Determination of Sulfur in Coal and Ash Slurry by High-Resolution Continuum Source Electrothermal Molecular Absorption Spectrometry." *Spectrochimica Acta Part B: Atomic Spectroscopy* 88, pp. 80–84. Doi: http://dx.doi.org/10.1016/j.sab.2013.04.011.

Niece, B.K.; and J.F. Hauri. April 2013. "Determination of Mercury in Fish: A Low-Cost Implementation of Cold-Vapor Atomic Absorbance for the Undergraduate Environmental Chemistry Laboratory." *Journal of Chemical Education* 90, no. 4, pp. 487–489. Doi: http://dx.doi.org/10.1021/ed300471w.

Nóbrega, J.A.; J. Rust; C.P. Calloway; and B.T. Jones. September 2004. "Use of Modifiers with Metal Atomizers in Electrothermal Atomic Absorption Spectrometry: A Short Review." *Spectrochimica Acta Part B: Atomic Spectroscopy* 59, no. 9, pp. 1337–1345. Doi: http://dx.doi.org/10.1016/j.sab.2004.06.003.

Ortner, H.M.; E. Bulska; U. Rohr; G. Schlemmer; S. Weinbruch; and B. Welz. December 2002. "Modifiers and Coatings in Graphite Furnace Atomic Absorption Spectrometry—Mechanisms of Action (A Tutorial Review)." *Spectrochimica Acta Part B: Atomic Spectroscopy* 57, no. 12, pp. 1835–1853. Doi: http://dx.doi.org/10.1016/s0584-8547(02)00140-4.

Pena-Pereira, F.; I. Lavilla; and C. Bendicho. January 2009. "Miniaturized Preconcentration Methods Based on Liquid–Liquid Extraction and Their Application in Inorganic Ultratrace Analysis and Speciation: A Review." *Spectrochimica Acta Part B: Atomic Spectroscopy* 64, no. 1, pp. 1–15. Doi: http://dx.doi.org/10.1016/j.sab.2008.10.042.

Pereiro, M.R.; A. López; M.E. Díaz; and A. Sanz-Medel. 1990. "On-Line Aluminium Preconcentration and Its Application to the Determination of the Metal in Dialysis Concentrates by Atomic Spectrometric Methods." *Journal of Analytical Atomic Spectrometry* 5, no. 1, pp. 15–19. Doi: http://dx.doi.org/10.1039/ja9900500015.

Petrak, J.; D. Myslivcova; P. Man; R. Cmejla; J. Cmejlova; D. Vyoral; M. Elleder; and C.D. Vulpe. February 2007. "Proteomic Analysis of Hepatic Iron Overload in Mice Suggests Dysregulation of Urea Cycle, Impairment of Fatty Acid Oxidation, and Changes in the Methylation Cycle." *AJP: Gastrointestinal and Liver Physiology* 292, no. 6, pp. G1490–G1498. Doi: http://dx.doi.org/10.1152/ajpgi.00455.2006.

Rapsomanikis, S. 1994. "Derivatization by Ethylation with Sodium Tetraethylborate for the Speciation of Metals and Organometallics in Environmental Samples. A Review." *Analyst*, 119, no. 7, pp. 1429–1439. Doi: http://dx.doi.org/10.1039/an9941901429.

Resano, R.; and E. García-Ruiz. January 2011. "High-Resolution Continuum Source Graphite Furnace Atomic Absorption Spectrometry: Is It As Good As It Sounds? A Critical Review." *Analytical and Bioanalytical Chemistry* 399, no. 1, pp. 323–330. Doi: http://dx.doi.org/10.1007/s00216-010-4105-x.

Rubio, R.; J. Albertí; J. Padró; and G. Rauret. July 1995. "On-Line Photolytic Decomposition for the Determination of Organoarsenic Compounds." *TrAC Trends in Analytical Chemistry* 16, no. 6, no. pp. 274–279. Doi: http://dx.doi.org/10.1016/0165-9936(95)91620-8.

Sanz-Medel, A. 1998. "Toxic Trace Metal Speciation: Importance and Tools for Environmental and Biological Analysis." *Pure and Applied Chemistry* 70, no. 12, pp. 2281–2285. Doi: http://dx.doi.org/10.1351/pac199870122281.

Sanz-Medel, A.; M.C. Valdés-Hevia y Temprano; N. Bordel García; and M.R. Fernández de la Campa. July 1995. "Generation of Cadmium Atoms at Room Temperature Using Vesicles and Its Application to Cadmium Determination by Cold Vapour Atomic Spectrometry." *Analytical Chemistry* 67, no. 13, pp. 2216–2223. Doi: http://dx.doi.org/10.1021/ac00109a048.

Shaltout, A.A.; I.N.B. Castilho; B. Welz; E. Carasek; I.B. Gonzaga Martens; and S.M.F. Cozzolino. September 2011. "Method Development and Optimization for the Determination of Selenium in Bean and Soil Samples Using Hydride Generation Electrothermal Atomic Absorption Spectrometry." *Talanta* 85, no. 3, pp. 1350–1356. Doi: http://dx.doi.org/10.1016/j.talanta.2011.06.015.

Síma, J.; P. Rychlovsky; and J. Dedina. January 2004. "The Efficiency of the Electrochemical Generation of Volatile Hydrides Studied by Radiometry and Atomic Absorption Spectrometry." *Spectrochimica Acta Part B: Atomic Spectroscopy* 59, no. 1, pp. 125–133. Doi: http://dx.doi.org/10.1016/j.sab.2003.11.005.

Slavin, W.; D.C. Manning; and G.R. Carnrick. 1981. "The Stabilized Temperature Platform Furnace." *Atomic Spectroscopy* 2, pp. 137–145.

Stafilov, T. July 2000. "Determination of Trace Elements in Minerals by Electrothermal Atomic Absorption Spectrometry." *Spectrochimica Acta Part B: Atomic Spectroscopy* 55, no. 7, pp. 893–906. Doi: http://dx.doi.org/10.1016/s0584-8547(00)00227-5.

Sturgeon, R.E. June 1996. "The Graphite Furnace and Its Role in Atomic Spectroscopy." *Analytical and Bioanalytical Chemistry* 355, no. 5–6, pp. 425–432. Doi: http://dx.doi.org/10.1007/s0021663550425.

Sturgeon, R.E.; and Z. Mester. August 2002. "Analytical Applications of Volatile Metal Derivatives." *Applied Spectroscopy* 56, no. 8, pp. 202A–213A. Doi: http://dx.doi.org/10.1366/000370202760249675.

Theodoroela, S.; N.S. Thomaidis; and E. Piperaki. August 2005. "Determination of Selenium in Human Milk by Electrothermal Atomic Absorption Spectrometry and Chemical Modification." *Analytica Chimica Acta* 547, no. 1, pp. 132–137. Doi: http://dx.doi.org/10.1016/j.aca.2005.01.030.

Timerbaev, A.R.; K. Pawlak; C. Gabbiani; and L. Messori. July 2011. "Recent Progress in the Application of Analytical Techniques to Anticancer Metallodrug Proteomics." *TrAC Trends in Analytical Chemistry* 30, no. 7, pp. 1120–1138. Doi: http://dx.doi.org/10.1016/j.trac.2011.03.007.

Tsalev, D.L. 1999. "Hyphenated Vapour Generation Atomic Absorption Spectrometric Techniques." *Journal of Analytical Atomic Spectrometry* 14, no. 2, pp. 147–162. Doi: http://dx.doi.org/10.1039/a807304j.

Tsalev, D.L. July 2000. "Vapor Generation or Electrothermal Atomic Absorption Spectrometry?—Both! *Spectrochimica Acta Part B: Atomic Spectroscopy* 55, no. 7, pp. 917–933. Doi: http://dx.doi.org/10.1016/s0584-8547(00)00202-0.

Vale, M.G.R.; N. Oleszczuk; and W.N.L. dos Santos. August 2006. "Current Status of Direct Solid Sampling for Electrothermal Atomic Absorption Spectrometry. A Critical Review of the Development Between 1995 and 2005." *Applied Spectroscopy Reviews* 41, no. 4, pp. 377–400. Doi: http://dx.doi.org/10.1080/05704920600726167.

Villa-Lojo, M.C.; E. Alonso-Rodríguez; P. López-Mahía; S. Muniategui-Lorenzo; and D. Prada-Rodríguez. 2002. "Coupled High Performance Liquid Chromatography—Microwave Digestion—Hydride Generation—Atomic Absorption Spectrometry for Inorganic and Organic Arsenic Speciation in Fish Tissue." *Talanta* 57, no. 4, pp. 741–750. Doi: http://dx.doi.org/10.1016/s0039-9140(02)00094-2.

Wang, Y.; X. Luo; J. Tang; X. Hu; Q. Xu; and C. Yang. 2012. "Extraction and Preconcentration of Trace Levels of Cobalt Using Functionalized Magnetic Nanoparticles in a Sequential Injection Lab-On-Valve System with Detection by Electrothermal Atomic Absorption Spectrometry." *Analytica Chimica Acta* 713, pp. 92–96. Doi: http://dx.doi.org/10.1016/j.aca.2011.11.022.

Welz, A.; H. Becker-Ross; S. Florek; U. Heitmann; and M.G.R. Vale. April 2003. "High-Resolution Continuum-Source Atomic Absorption Spectrometry—What Can We Expect?" *Journal of the Brazilian Chemical Society* 14, no. 2, pp. 220–229. Doi: http://dx.doi.org/10.1590/s0103-50532003000200007.

Welz, B.; S. Morés; E. Carasek; M.G.R. Vale; M. Okruss,; and H. Becker-Ross. September 2010. "High-Resolution Continuum Source Atomic and Molecular Absorption Spectrometry—A Review." *Applied Spectroscopy Reviews* 45, no. 5, pp. 327–354. Doi:http://dx.doi.org/10.1080/05704928.2010.483669.

Wolf, W.R.; and K.K. Stewart. July 1979. "Automated Multiple Flow Injection Analysis for Flame Atomic Absorption Spectrometry." *Analytical Chemistry* 51, no. 8, pp. 1201–1205. Doi: http://dx.doi.org/10.1021/ac50044a024.

Wu, P.; L. He; C. Zheng; X. Hou; and R.E. Sturgeon. 2010. "Applications of Chemical Vapor Generation in Non-Tetrahydroborate Media to Analytical Atomic Spectrometry." *Journal of Analytical Atomic Spectrometry* 25, no. 8, pp. 1217–1246. Doi: http://dx.doi.org/10.1039/c003483e.

Yebra, M.C.; and M.H. Bollaín. July 2010. "A Simple Indirect Automatic Method to Determine Total Iodine in Milk Products by Flame Atomic Absorption Spectrometry." *Talanta* 82, no. 2, pp. 828–833. Doi: http://dx.doi.org/10.1016/j.talanta.2010.05.067.

Yoza, N.; Y. Aoyagi; S. Obashi; and A. Tateda. December 1979. "Flow Injection System for Atomic Absorption Spectrometry." *Analytica Chimica Acta* 111, pp. 163–167. Doi: http://dx.doi.org/10.1016/s0003-2670(01)93258-1.

Zybin, A.; J. Koch; D.J. Butcher; and K. Niemax. September 2004. "Element-Selective Detection in Liquid and Gas Chromatography by Diode Laser Absorption Spectrometry." *Journal of Chromatography A* 1050, no. 1, pp. 35–44. Doi: http://dx.doi.org/10.1016/j.chroma.2004.05.078.

Zybin, A.; J. Koch; H.D. Wizemann; J. Franzke; and K. Niemax. January 2005. "Diode Laser Atomic Absorption Spectrometry. Review." *Spectrochimica Acta Part B: Atomic Spectroscopy* 60, no. 1, pp. 1–11. Doi: http://dx.doi.org/10.1016/j.sab.2004.10.001.

Index

THIS TITLE IS FROM OUR MATERIAL SCIENCE COLLECTION. OTHER TITLES OF INTEREST INCLUDE

Bonds and Bands in Semiconductors, Second Edition,
By J.C. Phillips

Plastics Technology Handbook - Volume 1:
Introduction, Properties, Fabrication, Processes,
By Don Rosato

Plastics Technology Handbook - Volume 2:
Manufacturing, Composites, Tooling, Auxiliaries,
By Don Rosato

Polymer Testing: New Instrumental Methods,
By Muralisrinivasan Subramanian

Solid-State NMR: Basic Principles & Practice,
By David C. Apperley

An Introduction to Transport Phenomena In Materials Engineering, Second edition,
By David Gaskell

Characterization of Tribological Materials, Second Edition,
By William A. Glaeser

X-Ray Fluorescence Spectrometry and Related Techniques: An Introduction,
By Eva Margui

Scattering of Acoustic and Electromagnetic Waves by Small Impedance
Bodies of Arbitrary Shapes,
By Alexander G. Ramm

Surface Engineering and Technology for Biomedical Implants,
By Yoshiki Oshida

Announcing Digital Content Crafted by Librarians

Momentum Press offers digital content as authoritative treatments of advanced engineering topics, by leaders in their fields. Hosted on ebrary, MP provides practitioners, researchers, faculty and students in engineering, science and industry with innovative electronic content in sensors and controls engineering, advanced energy engineering, manufacturing, and materials science. **Momentum Press offers library-friendly terms:**

- perpetual access for a one-time fee
- no subscriptions or access fees required
- unlimited concurrent usage permitted
- downloadable PDFs provided
- free MARC records included
- free trials

The **Momentum Press** digital library is very affordable, with no obligation to buy in future years.

For more information, please visit **www.momentumpress.net/library** or to set up a trial in the US, please contact **mpsales@globalepress.com**.